Positive Flying

Happy Flying!

Bill Gunther

Also by

RICHARD L. TAYLOR

INSTRUMENT FLYING
FAIR-WEATHER FLYING
UNDERSTANDING FLYING
IFR FOR VFR PILOTS

Richard L. Taylor
William M. Guinther

FOREWORD BY ARTHUR GODFREY

AN ELEANOR FRIEDE BOOK
MACMILLAN PUBLISHING COMPANY
New York

Macmillan Publishing Company
866 Third Avenue, New York, N. Y. 10022
Collier Macmillan Canada, Inc.

Library of Congress Cataloging in Publication Data

Taylor, Richard L.
Positive flying.

Reprint. Originally published: New York : Delacorte
Press, 1978.
"An Eleanor Friede book."
Includes index.
1. Airplanes—Piloting. I. Guinther, William M.
II. Title.
TL710.T362   1983   629.132′52   83-9812
ISBN 0-02-546570-8

First Macmillan Edition 1983

10 9 8 7 6 5 4 3 2

Macmillan books are available at special discounts for bulk purchases
for sales promotions, premiums, fund-raising, or educational use.
For details, contact:

Special Sales Director
Macmillan Publishing Company
866 Third Avenue
New York, N.Y. 10022

Designed by Oksana Kushnir

Printed in the United States of America

CONTENTS

FOREWORD

The first "how to fly" book I ever read was a dilly. Wish I could find it again. Written by some gutsy, self-taught pilot in about the mid-twenties, it listed some "don'ts," all of which I should have committed to memory. Some of them, in hindsight, were hilarious; but when I started to learn to fly in '29, they were taken very seriously.

"If the engine quits," said one, "land immediately!"

"But," it cautioned, "never land in a snowstorm!"

Since those days, I guess I've read every word that's ever been written in every book and magazine devoted to flying and learned something from each. I'm still reading and I'm still learning.

Mind you: I've been rather extraordinarily privileged over the years to have been checked out in nearly every make and model of conventional propeller, prop-jet, and pure jet airplane ever flown—from the Beech OX-5 Travelaire to the DC-10, with about ten hours in blimps, twelve hundred in helicopters, and mebbe twenty-five in gliders, to boot.

At least a hundred different test pilots and instructors, military and civilian, have worked me over—at my request usually—to keep me current and sharp.

And, after nearly seventeen thousand hours, I was just beginning to really feel "at home" in my airplanes, justified, I thought, in believing I had it "all figured out."

Then, this past summer (1977), I met Bill Guinther. In one two-hour session with him, I learned how little I really knew about flying.

Oh, sure, I can take you from here to there—been doing it safely for forty-eight years. Emergencies? Of course, lots of 'em: engines out, cockpit fires, bird strikes, lightning strikes, severe icing, electrical failures, gear problems. It's all happened. And I can't honestly say I always just "lucked out," either.

Some of it had to be lore and the refusal to panic— until afterward.

But a lot of it *was* just luck, because I knew nothing of the principles of Positive Flying, superbly spelled out for us in this book by Bill Guinther and Dick Taylor, the best-selling aviation author and educator who teaches in the same way.

I'll tell you how high I am on *Positive Flying:*

If I were the Administrator, this book would be the "bible" of the F.A.A. Every flight examiner would be required to read it, study it, master it, and then demonstrate an expertise to the satisfaction of Bill Guinther—personally! The examiners, in turn, would similarly require every flight instructor in America to demonstrate his or her competence in Positive Flying, thus *standardizing* the *teaching* of the art.

Since I'm now in my seventy-fifth year, however, I'll probably never get that F.A.A. job. Therefore, the best I can do, as one avid student pilot to another, is to beg you to read and master *Positive Flying.*

Let's all feel at home in the air!

ARTHUR GODFREY
December, 1977

INTRODUCTION

When the United States started to get its military machinery in gear for World War II, the training people recognized the need for an unprecedented number of airmen; thousands of pilots for the "heavies," thousands more for the fighters to protect them, and sad but inevitable, additional thousands to replace those who wouldn't come back. Repugnant perhaps from the humanistic point of view, the mass production of pilots was a military necessity, and training establishments appeared as though someone had scattered airport seed across the land. The name of the game was mini-

mum training time, give 'em something to build on,
wash out the ones who couldn't measure up, and be
sure that all square pilots would fit into round holes (or
vice versa) when they came out with shiny new wings
pinned to their uniforms.

This first American experience with mass pilot training proved effective in terms of standardization and
efficiency. There were individual conflicts, as the newly
minted pilots were distributed throughout the air
forces in accordance with the notorious "needs of the
service" (a true fighter pilot would sell his soul—or
whatever was required—to keep from being sentenced
to a bomber or a people-hauler). But the training meth-
ods *had* to produce pilots who could operate B-24s as
handily as P-38s; and a beginning cadet didn't know
whether he'd wind up fighting in a B-17, a P-51, or
hauling paratroops in a C-47.

Military expediency notwithstanding, this flight
training philosophy stood solidly rooted in the concept
that good flying habits had to be developed from the
outset. Those habits—"flying by the numbers"—were
expected to be refined and improved upon in the cruci-
ble of later experience, but the original aeronautical
experiences had to be the right ones; once a pilot went
to war, there were precious few second chances to
learn the fundamentals of the job.

The military educators in the early 1940s came up
with sets of rules (or "numbers") which would work for
all people in *all airplanes.* Obviously the numbers
which applied to a T-6 would mean nothing to a pilot
whose backside was strapped to a B-29, but the princi-

ple was the same: If he put a T-6 or a B-29 in a given attitude and applied a given amount of power, the "numbers" pilot *knew* that a certain performance would result. In other words, aircraft attitude combined with known power produces predictable performance. This concept, when applied to flight without external visual references, is known as "attitude instrument flying," and represents a very positive and disciplined method of making an airplane do precisely what the guy at the controls wants and expects it to do.

Attitude flying has carried over into airline operations. Most of today's captains cut their aviation eyeteeth on GI philosophies and find the attitude habit impossible to kick—the fact that it works might also have something to do with it. On top of that, even the smallest of contemporary air-carrier aircraft are so large that visual references and seat-of-the-pants flying are out of the question. Ask a four-striper, "How do you *fly* that big machine? You can't see the wingtips or even much of the nose," and he's likely to respond with something akin to "attitude plus power equals performance."

The bigger the airplane, the more important this piloting technique becomes; when a jumbo-jet takes off, the captain doesn't pull on the wheel until he "thinks" the nose is high enough, or until it "feels right"; he rotates the airplane to a precise, predetermined pitch attitude on the artificial horizon, and he knows that full power from the engines will produce the performance necessary for takeoff and climb. With experimentation and practice, a big-airplane pilot develops a set of attitude and power combinations which, with some fine tuning to accommodate density altitude and aircraft load, will produce predictable performance—performance he can count on.

In answer to the question, "How does one fly an airplane?" this method of attitude flying (it works just as well if outside visual references are available) really speaks more to the machine than to the man. The pilot's job is one of resource management; he manipulates a set of controls to set up and maintain a desired attitude, applies the proper amount of power, and the airplane flies nicely all by itself.

The formal education and aeronautical discipline (which is just a fancy way to say "flying by the numbers") of big-airplane pilots can be applied to small-aircraft operations; we can take the good things which have produced enviable safety records in airline and military aviation and use those principles when flying any kind of aircraft. And that's why we have written *Positive Flying*. Here is a method of training (or retraining, perhaps) which, for the first time in print, provides a cookbook procedure for How to Fly an Airplane.

A word of warning, however. It is a matter of applying the *principles* that big-airplane pilots use, not of blindly applying big-airplane techniques. As an example of the problems which can arise when the "whatever's good for General Motors is good for everybody" philosophy is used in flying, consider this unhappy story: Three airline pilots were going to work in their Apache when oil pressure on the right engine dropped to zero. Although the problem turned out to be a faulty gauge, the pilots knew that they'd be unable to feather the engine without oil pressure, and promptly shut down a perfectly good engine. (Can you imagine the "board meeting" that must have ensued in that cockpit, with three airline captains trying to decide what to do?) Obviously, the Apache could be flown on the remaining engine, so they set up full power, added carb heat (just in case), and promptly found that the airplane wouldn't maintain altitude. All three were Constellation pilots, and knew that partial flap extension provided additional lift on the triple-tailed transport, but there were no partial-flap markings on the Apache's indicator, so they decided to put down "some" flaps.

The story ends with a very well accomplished, low-damage, gear-up landing in an asparagus patch. The embarrassed trio did right in landing while still in control of the airplane, but they did wrong in applying a big-airplane technique to a little aircraft with a completely different type of wing. "Some" flaps might have helped the situation, but nowhere in the Apache book did that information appear, nor the proper airspeed to obtain the performance they needed.

Assuming that a pilot has learned to control airspeed, *Positive Flying* provides the airspeeds to control, the attitudes to go with them, and the power settings to obtain the desired performance. (The comprehensive Appendix at the back of the book contains that information for all the popular models of general aviation aircraft.) The emphasis is on reasonable performance in the normal situations, and on *safe* performance in the case of go-arounds and short/soft field operations. Study and practice with the numbers for your airplane, and you won't wind up trying to use Constellation techniques to fly the pride of Lockhaven.

IT'S NOTHING LIKE DRIVING A CAR

As an airplane can move in all three dimensions at once, most beginning pilots find themselves overwhelmed by the need to control left-right, up-down, and airspeed at the same time; after all, most of our vehicle-manipulating skills have been learned in ground-bound vehicles which move only in two dimensions, and rather solidly at that. There's a lot of flopping around the sky at the hands of pilots who have never learned the attitudes which produce the performance they desire; without something positive added to piloting, it's easy to become satisfied with airspeeds or altitudes that merely pass through the desired numbers once in a while.

Adding to the problem is the fact that each person perceives something different about what's happening as the airplane goes about its business; perhaps the most difficult task of the flight instructor is to help his student understand what he's supposed to see and what's supposed to happen in each maneuver and condition of flight. Often by dint of luck, a student pilot will react as though a light had been turned on in a dark room—"so *that's* what you meant!"—and the learning process goes ahead at a much more rapid pace. *Positive Flying* hopes to make uniform perception take place very early in the game by providing a series of definite attitudes and power settings which can be used from the outset; it hopes to replace luck with something you can sink your teeth into, something positive.

Positive Flying isn't the total answer to pilot training, but it is safe and certain—and it works. The secret? There is none; this method merely provides a set of proven attitudes, configurations, and power settings for the most frequently used conditions of flight, and suggests techniques for applying them. The dynamic nature of aviation means that Positive Flying will also probably fall short of answers to every operational situation, but if the method is used to develop basic habit patterns, it can easily be expanded to cover nearly anything a pilot might encounter.

Because it embraces a do-this, do-that philosophy, we believe that *Positive Flying* is going to put thinking back into general aviation—we'll supply the numbers, and you'll figure out how those numbers can do the most for you. There's a lot of room for innovation— these numbers are very conservative and designed with certain restrictions built in—but even the aviator who never ventures outside the Positive Flying approach will no doubt achieve outstanding fuel economy and engine life—and that can't be all bad!

Beyond the beginner, Positive Flying techniques find application in at least two other major areas of aviation: upgrading to a larger or more sophisticated airplane, and acquiring an instrument rating. Once again, in a situation where the pilot is faced with new experiences, new perceptions, it's comforting and efficient to rely on known, positive conditions. Especially during the quest for the IFR privilege, Positive Flying provides a safe foundation for learning how instrument flight is really handled; it's something to fall back on when the pressure builds (and don't think it won't!)—you can revert comfortably to numbers that you know will work.

Perhaps you'll chafe a bit at the premise that this technique will fly your airplane more smoothly and consistently than you can; but please don't let your sense of personal achievement be insulted, because the pilot who lets the machine do most of the work is a smart pilot—he uses the released time and attention to accomplish more meaningful tasks, such as looking for other airplanes, navigating, communicating, or the hundred other things he's supposed to do. And if in the bargain he turns out to be a safer pilot, so much the better.

If Positive Flying techniques can make aviation more efficient, economical, and pleasant when everything is running right, how about those unhappy times when problems show up? The virtues of positive action, backed by proven performance, can't be denied. To wit: Once upon a time, the pilot of a light twin-engine airplane tried to take off from an airstrip that was far too short. Past the point of being able to stop, and faced with an onrushing fence, he wanted in the worst way to get the airplane off the ground, using the book-

8

recommended short-field technique of lowering the flaps. But he hesitated, remembering that flaps create drag, and instead pulled the twin off the ground and staggered into the fence. The pilot died, in vain. If only he had known what his airplane, as loaded, would do, he could have fairly leaped into the sky and climbed at least 1000 feet per minute with full flaps extended! Certainly not a procedure we'd recommend for all flying machines, but the fact remains that a life would have been saved if that pilot had been aware of what his particular airplane would do in a tight situation.

You will probably find a whole herd of sacred cows put out to pasture as you go through *Positive Flying;* certain time-honored, super-conservative procedures and techniques replaced by real-world suggestions concerning How to Fly an Airplane—positively.

Perhaps this is the kind of book the airplane manufacturers would like to write about each model they build; manuals, operating handbooks, call them what you will, the "Here's Your New Airplane" publications contain plenty of information about the performance of the machine (how fast it will fly, how much runway is needed for takeoff and landing, how rapidly it will climb, and so on), but precious little about how to use it. Even the simplest kids' toys have explicit instructions: Put Tab A in Slot B, turn the key clockwise three times, push Button C, and a clown jumps out of the box every time. Most airplane handbooks just don't go far enough—the pilot must read between the lines and interpret what isn't there in order to make the machine do what was intended. Unfortunately, in most cases the

9

interpretation is founded on a pretty shaky concept and very little actual experience; the performance that results is at best inconsistent, at worst not good enough.

Positive Flying is a sincere effort to pass along to other pilots what we believe is a better, safer method of flying airplanes. Your authors were both trained this way, and have developed and refined the method in their day-to-day instructional activities. Almost a religion? You bet, because we know that Positive Flying works, and that must be the basis for anything in which one really believes.

So for the beginning flyer, we hope you'll find in *Positive Flying* a framework around which you can build a solid piloting technique; for those aviators at all stages beyond the beginning, now that you've got The Certificate, it's time to learn How to Fly an Airplane.

A
NUMBER OR TWO
TO START WITH

At the very core of the Positive Flying concept is a group of attitudes, airspeeds, and power settings—enough to see you through any flight situation, but not so many that you will have trouble learning them. In fact, after you've used them for a while, they will become second nature.

As with any aerodynamic exercise, the forces at work and the results they produce are closely linked, and careful consideration must be given to selecting the combination which will provide adequate performance and satisfactory response. The foundation numbers in applying Positive Flying techniques to your everyday flying are airspeeds and power settings. But for each type of airplane, the numbers are different; if you understand why, you'll do a better job of piloting. Putting first things first, come back to the drawing board and review the aerodynamic principle which underlies the whole method—pitch stability.

Surely you remember—it may have been yesterday or it may have been years ago—when a flight instructor, bent on gaining your confidence in the airplane, demonstrated its stability by means of inducing a phugoid oscillation. The word sounds like something out of a medical dictionary ("Sorry, but the doctor will have to remove your phugoid"), but it really refers to the oscillations of an aircraft as it attempts to maintain, or return to, straight and level flight. Does that refresh your memory a little? Remember how the instructor pushed the nose to one side, then took his feet off the rudder pedals and talked about directional stability while the airplane yawed back and forth and finally settled down where it started?

With some airplanes, the instructor would have demonstrated lateral stability by rolling the airplane gently to left or right, then releasing the controls and watching the wings come back to level, slowly but surely.

And then the big maneuver, the pièce de résistance, the one which proved that the airplane was stable in pitch. The instructor pulled the nose up well above the horizon, and carefully using rudder pressure to keep the nose on a point (asymmetrical thrust and all that), released the control wheel. Immediately the nose started down and, perhaps to your concern, kept on going down until it was well *below* the horizon. The airplane was now racing downhill, engine turning faster, airspeed building; then, like magic, the nose began to ease upward again, through the horizon and up to a point somewhat short of its initial excursion. Down again, though not so far, and back up, each oscillation a little less than the one before; finally, the airplane seemed to have had enough and returned to smooth, steady flight. This was not magic or luck or super-engineering, of course; pitch stability is simply the product of a basic aerodynamic principle.

To begin with, most airplanes are designed so that the center of lift (that infinitely small point through which all the supportive force of the wings is acting) is normally *behind* the center of gravity (an equally small and imaginary point at which the *weight* of the airplane appears to be concentrated).

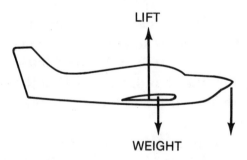

The nose-down force which results is partly offset by thrust—a safety factor of sorts because a *lack* of thrust (as in engine failure) would allow weight to assert itself and pitch the nose downward to maintain flying speed.

But thrust can't do the entire job. Furthermore, if it had to, every change in power would cause either a climb or a descent, and you'd have an airplane which would climb and descend at only one airspeed—whatever speed provided the proper relationship of thrust, lift, and weight. Enter the *horizontal stabilizer,* an upside-down airfoil (an airfoil is a structure which produces lift when air passes around it) placed at the rear end of the fuselage so that a tail-down force is generated to hold the nose level.

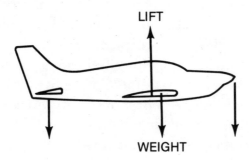

Add an elevator which can vary the downward force at the tail, and the pilot is able to keep the nose where he wants it. Extra thrust can be translated into more speed; on the other hand, by arresting the nose-down tendency when thrust is reduced, the pilot can fly at a *lower* airspeed.

Left to its own devices, and in consideration of the fact that an airfoil produces more lifting force as airspeed increases, this pitch-control system would ultimately find an airspeed at which the nose-down force is exactly balanced by the tail-down force of the horizontal stabilizer.

14

Let's say that for some hypothetical airplane this happens at 100 miles per hour, and at that airspeed the pilot decides to demonstrate pitch stability. When he pulls the nose up (no change in power setting, of course), airspeed begins to drop off and with it goes some of the down-lift exerted by the horizontal stabilizer. As soon as the pilot releases the controls the now nose-heavy plane pitches down and the airspeed builds; as the speed goes beyond 100 mph due to inertia, the nose-up force increases and starts the airplane uphill again. The up-down oscillations will continue, ever decreasing, until the airspeed stabilizes at 100 miles per hour—all the lifting forces having been brought into equilibrium.

The addition of an elevator trim tab provides aerodynamic muscle for the pilot when he wants the airplane to fly at any one of a wide range of airspeeds. When he discovers the angle of attack (directly related to pitch attitude in level flight) and power setting that will maintain altitude at a given airspeed, the trim tab merely "programs" the elevator aerodynamically for that speed; any airspeed greater than that for which it's programmed will make the elevator more effective and cause the nose to pitch up, while any less and the nose-down tendency will take over.

You would think that this stabilizing effect, this "airspeed programming," would work equally well at any speed from just-above-a-stall to redline—after all, more speed means more lift and vice versa—but in fact, the response is quite varied. At relatively *high* airspeeds, most aircraft tend to return to the original condition

15

very crisply and with little oscillation; at very low airspeeds, however, the tendency to stabilize is usually sluggish, sometimes the oscillations continue ad infinitum, and in extreme cases the ups and downs become divergent until the airplane either stalls or tries to perform an outside loop—something that is absolutely guaranteed to get your adrenaline pump running full bore! So, pitch stability and aerodynamic response become key items in the determination of the "proper" airspeed. In the interest of operational flexibility, *Positive Flying* provides two numbers—high-speed for enroute flight, low-speed for instrument approaches and VFR pattern work—both of which are well within the range of adequate aerodynamic response.

Power required vs. power available is next on the list of considerations. With the elevator programmed (trimmed) to maintain a certain speed, it follows that some specific amount of thrust will be required to hold the airplane in level flight. Additional thrust (a higher power setting) will cause the elevator to command a new pitch attitude, and of course that change will result in a climb. The opposite will be true of a decrease in power.

If the high-speed number were chosen with nothing but pitch response in mind, the resulting airspeed might be so high that it required every last horse under the hood—in which case the airplane could not possibly climb unless the airspeed changed. (Descent is never a problem; you can *always* reduce power.) Considering wear and tear on the engine, and the miserable fuel economy you'd experience by running at full throttle

16

all the time, the high-speed number should be a compromise between pitch response and the desirability of always having some power in reserve.

For the low-speed number, the compromise will trade some of the pitch stability (but not much) for more power available. To be useful and safe, the low-airspeed power requirement will always leave enough thrust available to generate a climb (at the programmed airspeed) of at least 500 feet per minute; that's considered sufficient to start moving you out of trouble in either a visual or instrument missed-approach situation.

In the case of airplanes with fixed-pitch propellers, the relatively low ratio of horsepower per pound of aircraft means that full power will probably be required to meet the 500 foot-per-minute criterion; as a result, the low-speed number will consider the fact that fixed-pitch RPMs are greatly dependent on airspeed. You don't want to be traveling so fast down final approach that the prop goes through the RPM redline when you open the throttle.

The limiting factor for airplanes equipped with constant-speed props is a little less severe, but in the interest of noise abatement and engine life, the suggested RPM will be a normal cruise setting, yet one that is high enough to safely absorb the manifold pressure required for a 500 foot-per-minute climb.

Maximum flap-extension speed is yet another consideration for those aircraft so equipped. Since flaps will be used only in final approach and landing situations, the high-speed number doesn't take V_{fe} into account.

(V_{fe} is the highest airspeed at which full flaps will stay fastened to the wings.) But the low-speed number, the one that will be used primarily for traffic patterns and instrument approaches, will always be below flap-extension speed so you can use the "special boards" the way they were intended—as *air brakes* when it's time to land.

Landing gear must also be considered in arriving at the airspeed numbers. If the undercarriage on your airplane is the type that stays put all the time, the drag of the wheels in the slipstream is one of the reasons the airplane won't go as fast as you wish it would. Given the power available for climb at either high or low airspeed, there's no way to alter the drag-thrust relationship.

But you who fly with retractable landing gear can do more with the wheels than simply move them up and down; you can use that drag (or lack thereof) to help fly the airplane. For example, suppose you've applied whatever power it takes to hold level flight at the low-speed number. When it's time to descend, lower the gear and down she goes; the *amount* she goes down will vary from one aircraft type to the next, but the addition of drag with no change in thrust has the same effect as reducing thrust, and we agreed earlier that whenever that happens to an airplane programmed to maintain one airspeed, the inevitable result is a descent.

Or how about this? You're flying level, gear extended, at the lower of your two airspeeds, and elect to execute a missed approach. When you retract the wheels, it's the same as increasing thrust, and the airplane responds by entering a comfortable, safe climb—at a rate proportional to the amount of drag that you just eliminated, *plus* the additional power that was needed to maintain level flight with the wheels down. Ah, the simplicity of it all. For retractable-gear aircraft then, the low-speed number will also take into consideration the airspeed limit for the landing gear.

In summary, Positive Flying uses two airspeeds—high and low, for en-route and approach operations, respectively—both of which have been determined after careful consideration of pitch response, power required for a safe climb, engine limitations, wing flap and landing gear extension speeds. Combined with the aircraft attitudes and power settings that result in the desired performance, you have a set of numbers that simplifies the mechanics and the thought processes of making your airplane do exactly what you want.

This takes away the feel of it all, you say? Makes flying mechanical? You bet it does, but don't be fooled by what you feel—the human "feel" system is loaded with illusions, and the pilot who depends on it will find that sense wanting sooner or later. Positive Flying closes the gate on that particular garden path, and provides instead a solid, dependable method that will work until the laws of aerodynamics are repealed.

NAILING
DOWN THE NUMBERS

The information and exercises in this chapter will help you develop, in your own airplane, the proper airspeeds, attitudes, and power settings for the three maneuvers fundamental to instrument approaches and VFR pattern work. The maneuvers are level flight, a 500 foot-per-minute climb, and a 500 foot-per-minute descent. For aircraft with retractable landing gear, the result will be three pitch attitudes and two power settings. Pilots of nonretractable types will need a third power setting to accommodate the descent.

Throughout the exercises, pitch attitudes will be referred to as "one nose-width high," or "one nose-width low," and so on. This description of what you see on the attitude indicator has been chosen as a compromise of all the attitude displays in the general aviation fleet. Most have a white dot in the center of the simulated airplane, and changing the pitch attitude up or down so as to move the dot one full diameter results in a "one nose-width" change. Some attitude indicators show only simulated airplane wings, with no "nose dot" at all; no matter, for in this case, "one nose-width" means one *wing*-width. Either way, the attitude indicator provides a scale on which to measure the pitch attitude of the real airplane. The bigger the indicator the better (why do you think the airlines and the military use attitude indicators six inches in diameter?), and should you be fortunate enough to have an instrument with a vertical scale marked in degrees, use that bonus for all it's worth; make your notations of pitch attitude in language such as "five degrees nose up," or "two-and-a-half degrees nose down."

The first step in providing your own numbers for Positive Flying is to "zero" the attitude indicator. On a smooth-as-glass day, trim the airplane for level flight at a normal cruise power setting; when everything settles down, move the adjustable airplane symbol until it rests precisely on the horizon line, covering it completely. If you haven't a pet power setting, try 65 percent; it works well with virtually every airplane. Turbocharged models are the exception; they produce uncommonly high airspeeds at their normal 75–80 percent cruise power settings. So you folks with turbo-

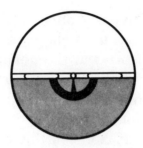

charged engines will have to throttle back to a rather plebeian 65 percent to make the method work.

Once the attitude indicator is adjusted to show zero pitch at normal cruise airspeed, don't change it; come back and land, and see what it looks like on the ground. Unless you're operating a Bonanza, Bellanca Viking, or a Cherokee Six, typical of those aircraft which tend toward a nose-high attitude on the ground, chances are that the little airplane will cover up the artificial horizon. But whatever the attitude is when you come back and land, that is the picture you should set up prior to each flight. A heavily loaded airplane will obviously sit in a bit of a squat, but don't change a thing; once airborne, you should fly the prescribed attitudes, and realize that some additional power will be required to compensate for the added weight.

There will be some determinations involved in arriving at the proper set of numbers, so it's a good idea to take someone along on the calibration ride. Aside from the clerical help and the pleasure of his or her company, an assistant gives you plenty of time to look outside for other airplanes.

22

UNCOMPLICATED AIRPLANES
—FIXED-GEAR, FIXED-PITCH PROPELLERS

Make your usual full-power takeoff, and as soon as possible after leaving the ground, put the airplane in a two nose-width high attitude. Hold it there, and trim away any elevator pressures until the airplane will continue climbing hands-off. Now, begin reducing power slowly and carefully until the rate of climb settles at exactly 500 feet per minute. You'll get the best results by keeping your hands off the control wheel, using whatever rudder pressure is required to keep the nose on a point. When the rate of climb settles at 500 feet per minute, have your clerk note airspeed and RPM on the worksheet on pages 24 and 25.

The next step is to smoothly and slowly reduce the power setting by 500 RPM; as thrust drops off, the airspeed-programmed elevator will pitch the nose down a bit in order to maintain the climb airspeed. Some models, notably the high-wing designs, will tend to hunt for the proper pitch attitude, but will eventually settle down in steady, level flight. The slower the power reduction, the fewer oscillations you'll experience. If it appears that a slight power adjustment is necessary to counter a climbing or descending tendency, go ahead; small changes in RPM to obtain absolute level flight are not uncommon. At this point, make

POSITIVE FLYING AIRSPEED DETERMINATION—
Fixed-Gear, Fixed-Pitch Prop

Step 1.
At full power, establish pitch attitude at two nose-widths UP; trim for hands-off flight. Let airspeed stabilize, then smoothly reduce power to obtain exactly 500 fpm UP.

	ATTITUDE	RATE OF CLIMB	AIRSPEED	RPM
INITIAL CLIMB	⬯	500↑	___	___

Step 2.
Smoothly reduce power by 500 RPM; airplane should fly level at same airspeed. Adjust power as necessary to obtain level flight.

	ATTITUDE	RATE OF CLIMB	AIRSPEED	RPM
LEVEL-OFF	___	↕	___	___

Step 3.
If airspeed in level flight changed by more than 10 mph from the two nose-width climb speed, repeat Step 1 at *one* nose-width UP.

	ATTITUDE	RATE OF CLIMB	AIRSPEED	RPM
SECOND CLIMB	⬯	500↑	___	___

SECOND
LEVEL-OFF ___ ↔ ___ ___

Step 4.

Repeat Step 1 again if necessary, at a different pitch attitude; you must determine the combination which will produce climb and level flight at ± 10 mph.

	ATTITUDE	RATE OF CLIMB	AIRSPEED	RPM
THIRD CLIMB	___	500↑	___	___
THIRD LEVEL-OFF	___	↔	___	___

Step 5.

From level-flight conditions determined in Step 2, 3, or 4, smoothly reduce power by 500 RPM; airplane should descend 500 fpm. Adjust power as necessary to obtain exactly 500 fpm DOWN.

	ATTITUDE	RATE OF DESCENT	AIRSPEED	RPM
DESCENT	___	500↓	___	___

a record of the pitch attitude, airspeed, and power setting.

> NOTE: Because of its aerodynamic characteristics, a high-wing aircraft will tend to change airspeed when power is changed. For example, check the figures for the Cessna 172 on Page 148 of the Appendix; notice that from a climb speed of 90 mph, the 172 levels off at 95 mph, and the 500 foot-per-minute descent takes place at 100 miles per hour. If your airplane has the wing on top, don't try to retrim, but accept this airspeed change as a normal condition. You will still obtain predictable performance, even though the airspeed varies from one maneuver to the next; and after all, who should apologize for maintaining airspeed within 5 or 10 miles per hour? You should be *bragging!*

On occasion, certain aircraft will exhibit a relatively large airspeed change at level-off. Should airspeed increase or decrease more than 10 mph from the trimmed climb speed, go back and try another climb, this time with a reduced pitch attitude—one nose-width—and see what happens.

When you arrive at a climb and level-off airspeed which remains consistent (except for the high-wingers), set up a descent by slowly and smoothly reducing power just enough to obtain a 500 fpm descent; for most aircraft, the power change will be in the neighborhood of 250 RPM. The reduction of thrust leaves the elevator no choice but to pitch the nose down to maintain the trim airspeed. Once again, a slight adjustment in RPM may be necessary to obtain exactly 500 feet per minute.

That's all there is to it; you've established a standard climb, level flight, and a standard descent at the same airspeed (or very nearly so—be satisfied if the airspeed needle doesn't move more than 10 mph during any of these maneuvers). You should also have noticed that changing rudder pressures were required to keep the heading from wandering; the nose tried to turn left when you applied power for the climb, and yawed to the right whenever you reduced thrust. Now that you know what will happen, anticipate it, and *make the nose stay put* with whatever rudder inputs are required.

MORE COMPLEX—FIXED-GEAR, CONSTANT-SPEED PROPELLERS

These airplanes will behave aerodynamically almost exactly like their brethren with one-piece propellors. In the matter of power settings, the language must be changed from RPMs to inches of manifold pressure, and as it works out, there is usually a close correlation; where the fixed-gear airplanes require approximately 250 RPM to make the difference between level flight and a 500 fpm descent or climb, those with constant-speed propellers will get the same results with a power change of about 5 inches of manifold pressure. In either situation, 50 RPMs or 1 inch of manifold pressure will usually change the vertical velocity by about 100 feet per minute.

For more detailed directions regarding a power setting from which to begin your determinations, see the section on retractables; you can't miss it if you'll just keep on reading.

MOST COMPLEX—RETRACTABLE-GEAR, CONSTANT-SPEED PROPELLERS

There's more for the pilot to cope with aboard these higher-performance steeds (the FAA even invented a special category, known as "complex" aircraft), so it becomes especially important to reduce as many actions as possible to strictly mechanical functions. Here's where the constant airspeed concept really shines, for if you change airspeed, you'll have to retrim the elevator—and that can't be done mechanically. But certain power settings and landing gear configurations are about as mechanical as you can get, which means that in times of stress you can make a couple of mechanical moves and *know* that desired performance will result.

Under normal conditions of weight and density altitude, you can run through the entire climb/level flight/descent routine and never exceed 65 percent of the engine's rated horsepower. With this in mind, select an RPM setting which will safely absorb that much power, and remember that if you need *more* thrust to handle a heavy airplane or a high density altitude situation, you'll have to increase RPMs so as not to overstress the engine. For most light aircraft, 2300–2400 RPM will do the job; but check the operating manual for your machine and be certain.

As soon as practical after takeoff, raise the gear and set the power handles for 65 percent. Trim the airplane for hands-off flight with the attitude indicator two nose-widths UP (assuming, of course, that you have previously zeroed it, as directed earlier in this chapter). When the airspeed stabilizes, slowly reduce manifold pressure until the vertical speed indicator reads exactly 500 feet per minute UP.

Slow, steady pressure on the throttle will do wonders for damping the pitch oscillations that may occur—you're in no hurry. Note the airspeed and manifold pressure.

When you've found the power setting that produces the desired climb rate, begin reducing manifold pressure (easy does it!) until the airplane assumes level

POSITIVE FLYING AIRSPEED DETERMINATION—
Retractable-Gear, Constant-Speed Prop

Step 1.

At 65 percent power, establish pitch attitude at two nose-widths UP; trim for hands-off flight. Let airspeed stabilize, then smoothly reduce manifold pressure to obtain exactly 500 fpm UP.

	ATTITUDE	RATE OF CLIMB	AIRSPEED	MANIFOLD PRESSURE
INITIAL CLIMB		500↑	_____	_____

Step 2.

Smoothly reduce manifold pressure by 5 inches; airplane should fly level at same airspeed. Adjust manifold power as necessary to obtain level flight.

	ATTITUDE	RATE OF CLIMB	AIRSPEED	MANIFOLD PRESSURE
LEVEL-OFF	_____	↕	_____	_____

Step 3.

If airspeed in level flight changed by more than 10mph from the two nose-width climb speed, repeat Step 1 at *one* nose-width UP.

	ATTITUDE	RATE OF CLIMB	AIRSPEED	MANIFOLD PRESSURE
SECOND CLIMB		500↑	_____	_____

SECOND LEVEL-OFF

ATTITUDE	RATE OF DESCENT	AIRSPEED	MANIFOLD PRESSURE
_____	↕	_____	_____

Step 4.

From level flight at the airspeed obtained in Step 2 or Step 3, lower the landing gear; the airplane should enter a 500 fpm descent.

	ATTITUDE	RATE OF DESCENT	AIRSPEED	MANIFOLD PRESSURE
DESCENT	_____	500▼	_____	_____

Step 5.

When descent is stabilized, leave gear down, and increase manifold pressure to the value determined in Step 2 or Step 3; the airplane should level off at the same airspeed. Adjust manifold pressure to obtain original airspeed.

	ATTITUDE	RATE OF CLIMB	AIRSPEED	MANIFOLD PRESSURE
LEVEL-OFF	_____	↕	_____	_____

Step 6.

Raise landing gear; airplane should enter a 500 fpm climb at the same airspeed.

	ATTITUDE	RATE OF CLIMB	AIRSPEED	MANIFOLD PRESSURE
MISSED-APPROACH CLIMB	_____	500▲	_____	_____

flight. Some small power adjustments may be needed to hold altitude, but more than likely the difference between climb power and level-flight power will be in the neighborhood of 5 inches of manifold pressure.

When everything settles down, record the pitch attitude, airspeed, and power setting. If the airspeed varies more than 10 mph either side of the climb speed, try again with a lower pitch attitude—therefore, a higher airspeed—in the climb. Experiment with a *one* nose-width high attitude and, if necessary, cut and try until you determine a combination of pitch and power that produces a consistent airspeed in the climb and at level-off.

Now you're ready to discover that "DOWN" on the landing-gear switch applies to the airplane as well as the wheels; in stabilized, level flight, lower the landing gear—don't touch anything else!—and watch the rate of descent ease down to 500 feet per minute. There may be a slight variance in airspeed in this configuration (once again, don't be concerned about changes of less than $\pm$ 10 mph), and you may have to sit through a couple of pitch oscillations. If necessary, adjust power slightly to obtain 500 feet per minute, and record the results: attitude, airspeed, manifold pressure.

Most of the retractable-gear airplanes today incorporate doors to close up the wheel wells and reduce drag. Some models are designed so that the doors open to let the landing gear extend, then close again to "clean up" the airframe; other doors remain open whenever the wheels are down. Either way, there will be a short period of time when the doors are in motion, and the additional drag may cause an unexpected and unwanted pitch change. Experiment with your airplane, and if there is a tendency for the nose to pitch up or down during the extend-retract cycle, apply whatever control-column pressure is necessary to overcome the attitude change. In almost every case, the airplane will eventually assume a "normal" pitch attitude after the gear is down and locked or fully retracted; but if it needs a little help in overcoming the door problem, don't hesitate to pitch right in. (Pun intended.)

While you've got the airplane stabilized in this configuration, smoothly add power until the manifold pressure gauge reads the same as it did during the initial climb. Like magic, the nose will pitch up a bit, the descent will stop, and the airplane will once again be in level flight at the same airspeed.

This usually occurs with a change of about 5 inches of manifold pressure, but whatever the numbers for your airplane, you've just proved that "x" inches of manifold pressure and the drag of the landing gear are completely interchangeable in producing or arresting a 500 foot-per-minute change of altitude.

It stands to reason that if the airplane will fly level, landing gear extended, at the same power setting which produced a 500 foot-per-minute climb with the gear *retracted,* you'd need only to pull up the wheels to enter a climb—and that's exactly the way it works. When you eliminate the drag of the extended landing gear, the airplane senses an effective increase in thrust, and counters the new condition by pitching up to the same attitude you established in the initial climb. It's so easy and logical; you may even be so impressed you'll want to start referring to the landing gear control as the "missed-approach switch." (If you noticed an unusual pitch change due to the wheel-well doors when you extended the gear, be ready for a similar response on retraction.)

WHERE DO WE GO FROM HERE?

The numbers derived from the exercises you've just gone through must be regarded as starting points—foundations which will produce aircraft performance very close to what's desired. But the power setting and attitude for a 500 foot-per-minute climb may be slightly different today than it was yesterday, and the culprit will likely be increased density altitude, or a heavier load on board the airplane. While yesterday's attitudes and power settings will produce the same *airspeed,* the performance numbers will show the inevitable effect of thinner air or more weight. Under these circumstances, don't expect your airplane to climb quite as rapidly, be prepared for a greater rate of descent, and know that you will need to increase the level-flight power setting a bit to keep the altimeter where you want it. In other words, the foundation power settings may have to be increased across the board to obtain "standard" performance. How much? The answer to that is easy—whatever power is required to get the job done.

But let's assume that sometime after determining

35

your own performance numbers, you find that even when conditions are the same, the old faithful power settings aren't quite cutting the mustard—the throttle setting which used to produce level flight now results in a slow descent, you can't get the 500 foot-per-minute climb you've gotten used to, and when you throttle back or lower the gear for a standard descent, the vertical velocity indicator goes well beyond the 500 mark. Given the predictability of aircraft performance when using the Positive Flying method, there is only one thing to suspect—you've got an engine problem.

All reciprocating engines begin to lose some of their stuff from the very first time they are started up, and a meticulous record of power settings versus performance over the life of the engine would show a very slow, but inexorable, deterioration of powerplant output. But the change you need to watch for is the relatively *sudden* change; perhaps an unexplainable increase in power required since the last time you flew. When the load is the same and the density altitude not all that much different, a wise pilot will treat such an indication as a red flag warning that something unpleasant is getting ready to happen, and he'll get the airplane back on the ground in short order. How much better to be safely on the ground, wishing you were flying, than to be aloft with a just-quit engine and wishing you were on the ground.

IFR
APPLICATIONS

Flight in instrument conditions is perhaps the most dramatic demonstration of Positive Flying. The workload imposed on any pilot is bound to increase when he's "on instruments" because of the more precise navigation required, the additional communications responsibilities, and—let's face it—preoccupation with the primary task of keeping the airplane right side up. All this becomes even more demanding as you get to the most critical part of an IFR flight, the approach. Positive Flying will relieve much of that extra workload, for when the aircraft-handling chores are relegated to proven, reliable attitudes and power settings, you can have confidence that the airplane will keep right on doing what you want it to, and you will have more time to navigate and communicate.

The purpose of this chapter needs to be qualified a bit. Since there are almost never two IFR trips exactly the same from an operational standpoint, what we've written is not intended to cover the totality of IFR operations. Rather, we've placed the emphasis on the basics of handling your aircraft in instrument conditions; approach procedures, after all, are merely combinations of turns, climbs, and descents—the simple, basic maneuvers of flight.

MAKE SURE EVERYTHING'S WORKING

The most important item that must be accomplished before taking any airplane into IFR conditions is a complete pre-takeoff checklist, one which guarantees that all the instruments on which you're about to stake your well-being are functioning properly. The usual VFR once-around-the-panel-and-go checklist is just not good enough for instrument operations; if something goes wrong on a clear day, it's a fairly simple matter to point the airplane toward the nearest airport, descend, and land with visual attitude control and the power settings developed earlier. But when you can't see outside, the situation becomes nearly untenable, and you'll wish—

oh, how you'll wish—that you were on the ground instead of in the air.

A good IFR checklist should be designed as such: a thorough, systematic way of making sure that all the gauges are working. If you fly the same airplane all the time (that's a boon to proficiency in itself), design your own checklist to start at one corner of the instrument panel and proceed in an orderly fashion until you wind up at the other end with everything checked. For the most part, the checklist published by the manufacturer provides a good platform from which to begin, amended and supplemented with whatever additional checks and tests will provide you with the knowledge that all is well.

A complete check of the systems which power the gyroscopic flight instruments is absolutely essential. Nearly all of today's airplanes are equipped with redundant power systems—the directional gyro and attitude indicator are vacuum-powered, while an electric source is provided for the turn-and-bank indicator— but remember that vacuum pumps are engine-driven, and, especially on single-engine airplanes, a sheared shaft, an engine that won't turn up full RPM, a faulty vacuum pump, a broken or leaking vacuum line, puts you suddenly in a partial-panel situation. While checking the vacuum system for proper readings and indications, be certain the electric turn-and-bank is doing everything *it's* supposed to—it could get you out of the clouds in one piece someday.

In any event, design an IFR checklist with your

equipment and needs in mind, and *use* it—before
every flight.

TAKEOFF AND CLIMB

If for no other reason than the fact that you can't see
outside, it's good practice to leave the engine at full
power longer during an IFR takeoff. Safety can be translated into feet above the ground, and one of the best
ways to guarantee that all-important clearance is to
climb—at least initially—with all the horses your engine can get together. Even if there's a time limitation
on maximum power (a few general aviation power-
plants are *not* certified to run wide open forever), don't
be in a hurry to reduce power in an instrument climb;
leave all the go-handles forward until you feel comfortable. It's quite possible that you will feel more comfortable at progressively lower altitudes with experience,
as you develop a real awareness of the proper attitude
for the most efficient climb speed.

While there should be no rush to ease off on the
power setting, you should see that the airplane accelerates to its best climb speed as soon as possible. With the
climb attitude established and full power applied,
you're doing everything possible to put more distance
between you and the ground.

The next few minutes are going to be filled with
things to do other than fly the airplane. In even the

least complex IFR environment, you'll have some communicating to do, vectors to follow, perhaps an airway or VOR radial to intercept; and around the bigger terminals, the single pilot soon develops the cat-on-a-hot-tin-roof syndrome. So why not trim the airplane as soon as climb speed is achieved, and let the airplane do most of the work? Trim away control pressures whenever they become evident, and between radio transmissions and transponder squawks and the like, continue trimming until the airplane is truly flying hands-off. When you get to this point, passengers should be advised to sit still and shut up, because sliding a seat back or leaning forward to say something may affect rate of climb by as much as 15 feet per minute—friends, that's *trimmed!* (We know one pilot who trims so well that he's afraid to move his head once he gets things set up.)

Sooner or later, an IFR departure will require a heading change, and a sacred cow we'd like to butcher right away is the standard-rate turn. Great for timed turns in a partial-panel situation, and equally outstanding when you're jumping through the hoops of an IFR checkride, the 3-degree-per-second turn is not recommended while climbing. Because of the relatively low airspeed in an IFR climb, 10 degrees of bank provides a rate of turn which is adequate for this phase of flight. For example, at 100 knots, a 10-degree bank will cause the heading to change at the rate of almost 2 degrees per second; plenty fast to accommodate the turns you'll be asked to accomplish. Now and then you'll be asked to make a heading change of 180 degrees, but most climbing turns result in heading changes of 30, 45, or 60

41

degrees—and the difference in time-to-turn between the standard rate and 2 degrees per second becomes academic. Have you ever heard a controller ask a pilot to speed up his rate of turn?

Perhaps more important, a 10-degree bank can be accomplished smoothly and accurately with rudder pressure alone—no need to risk overcontrolling by leaning into the ailerons and disturbing the airplane in its least stable axis. A shallow bank will also take very little vertical lift from the wings, so little that climb performance will not be noticeably affected; most light airplanes can actually be trimmed for climb speed in such a turn. If you're still not convinced, consider this: A 10-degree bank, while adequate to get the job done, allows the heading change to proceed at a slightly slower pace, and gives the pilot a better opportunity to stay ahead of the airplane. There will be fewer instances of overshooting an assigned heading, with the bother of having to turn the other way to get back to where you should be (which of course introduces yet another chance for an overshoot, ad infinitum).

LEVEL-OFF AND CRUISE

Based on your determinations from "Nailing Down the Numbers" (Chapter 2), the process of leveling off will consist of reducing power slowly and smoothly until the altimeter stands still. The resulting cruise speed will be the same as climb speed, of course, and there's nothing wrong with that when you're beginning in the IFR business—one trim setting takes care of all situations.

But more than likely, that airspeed will be lower than you'd like at cruise, so the elevator must be retrimmed. The new trim setting will provide the aerodynamic foundation for everything that takes place from now until the instrument approach begins at the other end of the trip, so it's important that the level-off be accomplished properly, and that the matter of trimming be placed at the head of the list of things to do. Radio calls, precise navigation, and such should take second place; you must fly the airplane first. ("Stand by one" may have to be your reply to a controller at this stage of the game. Answer as soon as possible, but not before you have your house in order.) A pilot who allows his attention to be divided may find himself fighting the airplane the rest of the flight, all because he didn't take time to trim properly at level-off.

When climb power is to be maintained for cruise, a few practice sessions will develop your ability to accomplish the level-off with elevator trim alone. The objective is to blend a change in pitch attitude with an increase in airspeed and a reduction in climb rate; the *trick* is to make all three of these happen simultaneously at the desired altitude. A good-enough rule of thumb for light airplanes is to begin rolling in nose-down trim when the airplane is about 50 feet under the desired altitude. (This will work reasonably well for airplanes which climb at approximately 500 feet per minute; for more precise numbers, you may want to start your experimenting at an altitude which represents 10 percent of the actual climb rate, *i.e.*, 20 feet for a climb rate of 200 fpm, 100 feet for a climb rate of 1000 fpm, and so on.)

Patience is the watchword as you trim; make a small adjustment, then wait to see what effect it has. It's easy to count the turns of a trim crank, and most trim wheels have knurled edges or some sort of bumps on the rim; you can use these built-in markers to let you know when you've made small changes. Make one bump or a quarter-turn at a time (with an electric trim switch, one "blip" at a time), then see what happens; then a little more, until the pitch attitude is where you want it and the climb has stopped. If you don't care to level off with trim (it's a free country), ease the attitude indicator to its new position with elevator pressure, then slowly trim off the pressures that result. Either way, the trimming process should be complete in no more than a couple of minutes, and you can then devote your attention to the less important chores at hand. ("Okay, Center, now what was it you wanted?")

There's always the fellow who claims, "I like to leave a little bit of pressure on the wheel—leave the airplane just a little out of trim so I can 'feel' it." That's pure hogwash, because the human muscular system isn't built to function that way. Sooner or later—usually sooner than you'd think—your muscles will relax a bit even though you don't realize it, and if the airplane is trimmed for a higher or lower airspeed it will seek that speed, losing or gaining altitude in the process. And if such a pilot *is* able to make his airplane fly at a constant altitude, he's expending a great amount of energy and concentration on a task which should be relegated to the trim system.

The additional airspeed which results from maintaining climb power at level-off is the "high cruise" situations; "high" when compared to the lower speed recommended for instrument approach work. There's a good chance, however, that your airplane is capable of considerably higher cruise speeds; and of course you'd like to utilize the additional speed to justify the investment in the airplane. But for the first twenty hours—give or take a few—of your IFR experience, it's best to lope along at a lower airspeed until you get accustomed to the airplane and the speed at which things happen in the IFR system; why rush into trouble? The combination of a new, fast airplane and a new, not-so-swift IFR pilot often reads like something out of a Dick and Jane book: "See Dick's new airplane? See Dick's new airplane go fast? See Dick trying to catch up with the airplane? See the airplane running away with Dick?"

Not only does a fast flying machine frequently outspeed its pilot's thinking (a safety consideration), there's also a fallacy that going fast means saving a lot of time. For example, two pilots in identical airplanes take off and climb to the same cruise altitude, each using ten minutes in the climb; at level-off, one pilot sets up 140 knots, the other adds enough power to produce 160 knots. If both fly for 200 miles, the hare will arrive just eleven minutes ahead of the tortoise.

The extra time—at least in the beginning stages of your IFR experience—can be better spent getting accustomed to the airplane, the ATC system, and yourself—to say nothing of the fuel conservation inherent in the lower power setting.

DESCENT

The airspeed which satisfied your needs at cruise can be continued throughout the IFR letdown, with two major advantages: You won't have to shift mental gears a lot when estimating time-of-arrival (wind speed generally drops off as you descend, but the effect on rough estimates is usually not significant), and you won't have to make any trim change until leveling off at the new assigned altitude. The technique, then, is to *reduce power* when cleared to descend; if you've done a proper job of trimming, the airplane will nose over a bit of its own accord, and start downhill at a rate determined by the amount of power you take away.

Be certain to reduce power enough to establish an honest 500 foot-per-minute descent. The airplane will be somewhat lighter because of fuel consumption, and the power setting which worked so well during practice sessions may not do the job. It's best to start with the old familiar target power settings, then adjust as necessary to achieve 500 feet per minute.

With two quantities—airspeed and vertical speed—

held constant, it's easy to figure how far out you should request a descent; at 500 fpm, the altitude to be lost times two equals the time required for descent, and groundspeed will provide the distance required.* In all but the really busy terminal areas, controllers will usually issue a clearance for a lower altitude upon request.

For those occasions when you need to descend considerably faster than 500 feet per minute (when Approach Control is unable to let you down early, or when you'd like to stay high in a brisk tailwind as long as possible, or when you're flying above an ice-filled cloud deck and don't want to penetrate it any sooner than you have to), Positive Flying offers a solution: Lower the landing gear. If your descent airspeed is higher than the gear-limit speed, trim nose-up just enough to slow to the limiting speed, drop the rollers, and watch the rate of descent increase due to the additional drag.

LEVEL-OFF

Although jet operations are moving more and more to the concept of profile descents, a procedure which entails a continued loss of altitude at relatively high speed right up to the approach fix, most low-perfor-

---

*Think in terms of thousands. For example, cruising at 9000 feet, airport elevation 1000 feet, you must descend 8000 feet, which will require sixteen minutes at 500 fpm (8 X 2 = 16). So if groundspeed is 150 knots, you should request descent 40 miles out.

47

mance airplanes will be assigned an altitude which must be maintained for a longer period of time. When level-off time arrives (assuming that the descent has been flown at high cruise airspeed and 500 feet per minute), leave the power at the descent setting (which may have to be adjusted downward a bit due to the increasing air pressure at lower altitudes), and trim for your airplane's low cruise airspeed—it's the last trim setting you'll have to make for the balance of the approach. Make whatever fine power adjustments are necessary to hold level flight at the new altitude.

When the approach is being flown in a radar environment, it's quite likely that the controller will ask you to maintain a higher-than-normal airspeed right up to the approach fix—a request that should be honored *if you feel up to it.* On the other hand, an approach to a nonradar airport, or a completely uncontrolled terminal, often results in an extended time at the level-off altitude while you execute a procedure turn or fly a vector to intercept the final approach course. Excess speed—above that which you worked so hard to develop for IFR maneuvering—is completely unnecessary in this situation. Whether you are outbound for a procedure turn, or inbound after completing it, or just flying up to the final approach course, excess speed serves only to lengthen the approach, carry you farther away from the airport, use more fuel, and most important, make everything happen faster. Even the sorriest of duffers knows that you've got to keep your eye on the ball in golf, and an axiom has grown up in that sport

—"Want to see a bad shot? Look up." The same philosophy can be applied to instrument flying: "Want to fly a bad approach? Speed up." Don't let it happen to you. Slow down.

FINAL APPROACH

Every approach procedure includes a point at which descent to the minimum altitude begins. Using the ILS procedure, this is the point at which you intercept the glide slope; with some NDB and VOR procedures, the point is determined by station passage; in others, where the radio facility is located right on the airport, descent commences at the completion of the procedure turn. In any event, you should fly up to that point at your "working" airspeed, with the airplane well trimmed and with just enough power to maintain level flight.

In a retractable-gear airplane, start the descent by lowering the wheels—the added drag will do the job. When you're flying a fixed-gear machine, reduce power to the setting that will let you down at 500 feet per minute. These are *target* settings and may need to be adjusted for the situation at hand, but they are a place to begin.

A pilot-helper on some ILS approach procedures, though one that's seldom used, is the fact that the procedure-turn altitude is usually several hundred feet higher than the glide-slope intercept altitude. Use this difference to your advantage by staying at the higher altitude until the glide-slope needle centers—even if the controller has cleared you for the approach, which would allow you to descend legally to the intercept altitude. By staying at the procedure-turn altitude, you will inevitably intercept the glide slope earlier, and the extra minute or so (depending on groundspeed) provides a beautiful opportunity to get the rate of descent stabilized. As wind speed and direction change throughout the rest of the approach, you'll have to make the appropriate corrections to both heading and power setting, but who can argue against using that extra time to start getting all your descent ducks in a row?

In low-performance airplanes, it's unlikely that a rate of descent much greater than 500 feet per minute will be required either to stay on the glide slope of an ILS, or to make it to the MDA (Minimum Descent Altitude) of a nonprecision approach before the airport appears out of the mist. If it becomes apparent that a greater rate of descent *is* needed, make a slight power reduction—a couple hundred RPMs, an inch or two of manifold pressure, whatever is required. You'll soon discover that there's plenty of room left for good pilot judgment.

At the MDA on a nonprecision or circling approach, the primary concern is to stop the descent. In either type of airplane—fixed-gear or retractable—add power, whatever it takes to maintain level flight. A word of caution to pilots of retractables: Do not, repeat DO NOT, under any circumstances raise the landing gear to level off. Oh, it works—the descent will stop. But you've just set yourself up for one of aviation's classics, the gear-up landing. It will likely be the smoothest landing of your career, but you'll need full power to taxi.

LANDING

If there's one situation in flying where the human visual system is most likely to fail, it's at the end of an instrument approach. Experience continues to prove— even to the pros—that there are more illusions and distracting visual inputs between breakout and touchdown than at any other time. (And of course, you're close to the ground, where each error in judgment multiplies the chances of something unpleasant taking place.) The major troublemaker is foreshortening of the apparent horizon by fog, haze, low clouds—whatever has clanked up the atmosphere so that you don't see the normal picture through the windscreen. There's a strong temptation to push the nose of the airplane over so that what is actually seen looks the way it should—and when the perceived horizon is much closer to the airplane than on a normal-visibility day,

51

you've just fallen victim to the notorious "duck-under" maneuver.

Do you think you will be able to resist that temptation when it shows up some dark night, or some foggy day? The duck-under is such an insidious trap that airline safety experts believe in fully-coupled autopilot approaches (that's where the pilot is an "interested observer") to as low an altitude as possible—and if it's tough for airline captains to overcome this illusion, the nonprofessional pilot should treat a low-visibility approach with double caution.

Fortunately, there's an easy way to defeat the duck-under situation; remember that the *artificial* horizon, the one displayed on the attitude indicator, will stay put regardless of outside conditions. So look outside to see where you are in relation to the airport, but *disregard the real horizon and rivet your attention to the attitude indicator.* You've done everything right so far —why change?

There are other techniques which can be used during the landing phase of an approach to reduce the possibility of being led astray by what you *think* you see. One concerns the use of wing flaps. We strongly recommend that you refrain from extending them until the runway is in sight, and never until you're within 500 feet of the ground. In a low-visibility environment, the pitch change as the flaps come down (true of most airplanes) can do strange things to visual inputs, none of them good; so, when you extend the wing flaps, apply whatever pressure to the control column as required to keep the pitch attitude constant. The airplane will slow down, of course, but more important, the pitch "picture" won't change. There is no need to look for the flap switch or handle—it's right where it's always been, and taking your eyes from the runway can do nothing but introduce more visual problems. And as the elevator pressure builds up, get busy on the trim control to overcome the push. Trim, trim, trim—it's the secret to success.

In a weather situation where the time from breakout to touchdown is very short, you may be asking your eyeballs to adjust to changing inputs faster than they are capable—at least faster than the usual VFR approach and landing where there is plenty of time to judge rate of descent, airspeed, runway closure rate, and so on. With fewer visual cues, and less time in which to assimilate and interpret them, you may have to settle for a more-or-less mechanical landing: Extend the flaps, reduce power smoothly to idle, bring the nose up to the horizon, and wait for touchdown to happen. Care must be used in determining the nose-horizon relationship if you sit unusually tall (or short) in the saddle; know in advance what your personal landing picture should be.

Safely on the runway, why risk ruining a perfectly good approach and landing by inadvertently raising the landing gear when you thought you had grabbed the flap handle? Don't touch anything—flaps, cowl flaps, radios, carb heat, whatever—until you have cleared the runway and can take time to be certain what you are doing.

MISSED APPROACH

If there ever was an inflight situation designed to frustrate everything an instrument pilot is trying to accomplish, it's the missed approach. You've worked hard to get down close to the runway, but when the concrete is only a couple hundred feet below, the whole caper has to be called off—more than likely by an instructor or check pilot, but now and then because of weather. (If the atmospheric situation is so bad that there's a good chance a missed approach will be necessary, maybe you shouldn't be there at all.) In any event, the missed-approach maneuver has but one function— to start the airplane climbing—and with Positive Flying, it turns into a real pussy-cat exercise.

With a fixed-gear airplane, missing an approach means adding power until the tachometer registers the proper number of RPMs for your previously determined climb. Press the throttle forward smoothly, give the elevator a little help in damping the pitch oscillation if necessary, and trim as required; *then* attend to your other piloting chores. For instance, let the guys in the control tower know what you've done, and what you intend to do next. ("Tower, 23 Bravo, missed approach." "Roger, 23 Bravo, what are your intentions?" "This is 23 Bravo, what are my options?") *Don't trust what you feel*—you proved earlier that the airplane would climb super-safely in this condition, so don't upset things by trying to take over manually.

The missed-approach maneuver for retractables is

not much different: Add power (up to the predetermined setting, which in itself will stop the descent), then raise the landing gear and watch the climb begin. No muss, no fuss, no bother—and it's *safe*. Once again, be aware that there may be some strange physiological and psychological forces at work during the missed approach, so *don't trust your sense of feel!* Set up a condition of attitude and power that you know will result in a climb, trim as necessary, and let the airplane take care of itself.

Fixed gear or retractable, if you've done your homework properly, the numbers you need—airspeed and rate of climb—will result when you set climb power. Remember that a missed approach is just another maneuver in your bag of piloting tricks; it's not an emergency procedure, nor an exercise to find out if the engine can develop maximum horsepower.

CIRCLING APPROACH

At airports not blessed with ILS or straight-in VOR or NDB approaches, the airplane must be flown in some sort of pattern to arrive at the landing end of the runway. In nearly all cases, such maneuvering is required because the final approach course from the radio navigation aid does not line up with any of the runways; but now and then a terminal is situated so close to high terrain or other obstructions that the required vertical clearance causes you to arrive on final at an altitude much too high for a safe descent. There's little choice but to circle while losing altitude.

A circling approach demands the utmost in pilot skill and judgment, for it requires flying around the airport in low-visibility conditions, at an altitude considerably lower than that used in a normal landing pattern. If for no other reason, the sudden transition from all-instrument clues to all-visual (outside) inputs can impose extremely heavy loads on the eyeball-brain system.

The most important task at this point is to maintain a safe airspeed (you should be flying at the low cruise airspeed determined in Chapter 1) at an altitude no lower than the MDA specified for circling, and the only way to make sure this happens is to add power as required. The level-flight, low cruise power number will provide a safe margin above stall, even if you need to bank rather steeply to keep the field in sight and fly around to the landing runway (down this low, you should consider a 30-degree bank *steep*).

Pretend that every circling approach is being conducted at night, over rough terrain in unfamiliar country, and use no flaps until you're established on a solid, comfortable final approach. The rules call for the circling MDA to be maintained until it is necessary to go lower in order to land (this MDA guarantees only 300 feet of obstacle clearance), and there's nothing wrong with putting off flap extension until the last minute. As a matter of fact, you can inject a huge dose of safety into the circling situation by adding 800 or 1000 feet to the airport elevation (the normal VFR pattern altitude, one your eyeballs are accustomed to) and treating that as your personal minimum; if the airport is not in sight when you reach that level, continue down to the published circling MDA, but consider that altitude a *minimum* minimum. The published circling MDA must be treated with great respect; it's your guarantee of safety.

One more safety suggestion: If you ever find yourself having to bank more than 30 degrees to get lined up with a runway during a circling approach, consider that you've blown it. Go around and try again. The next approach should be much safer because of what you learned the first time. Don't try to sneak in; the famous last words, "It's my home base," are often just that— famous *last* words.

FLYING INSTRUMENTS IN ICE AND TURBULENCE

The traditional advice, of course, is *don't!* But if you fly IFR enough to make the rating worthwhile, and if you fly IFR in those parts of the country which are blessed (cursed?) with seasonal weather changes, the chances are very good that sooner or later you will find ice sticking to your airplane, or you will fly close enough to a thunderbumper to feel its effects. So, what should be done if that unhappy situation does occur? "Don't do it" doesn't help a bit when you're icing up, or about to be swallowed by a CB, so here are some techniques which can help get you out the other side in one piece.

ICING

There seems to be no doubt among those who have been there that the best way to combat structural icing is to get out of it—as soon as possible. Icing conditions usually cover relatively wide areas laterally, but are seldom very thick in the vertical dimension, so the most effective direction in which to move is up. (However, there will be times when "up" is limited by lack of oxygen or power. And in other situations, when you *know* that there is warm air below and terrain will permit, descending is the far better thing to do.) Timing is very important, especially with low-performance airplanes; but if you begin climbing within, say, ten minutes after noticing ice, and can climb at least 500 feet per minute, chances are good that you'll escape the sharpest claws of the ice monster. Power is the secret, so crowd on all your engine will stand; in addition to providing the thrust needed for climb, a high RPM tends to prevent ice accumulation on the propeller blades. (Go all the way to redline; most engines will carry maximum RPM all day—but check your powerplant's restrictions just to be sure.) And while pre-flighting for an IFR flight where icing is a possibility, polish the leading edges of the prop blades—a shiny surface is bound to collect less ice.

Assuming that you are flying at high cruise airspeed when the icing is encountered, make a rapid and positive transition from level flight to climb. Full power first, then raise the nose to climb attitude and let the airspeed settle down of its own accord. There will probably be an immediate "free" 1000 fpm rate of climb because of the zoom effect, and that's good—while the airspeed is dissipating, you're climbing, which is the name of the game at this point. Trim for the new airspeed, and continue climbing to the highest altitude you (or the airplane) can tolerate. Keep the bad weather below. By the way, an oxygen mask may well be the most effective anti-icing system of them all.

With Positive Flying numbers to work with, it's possible to rate the aircraft's performance and predict just how long you can tolerate a given level of ice accumulation, and whether you can climb out of trouble. For example, suppose that the first fifteen minutes in icing conditions require an additional 2 inches of manifold pressure or a couple hundred RPM just to maintain the status quo in terms of airspeed and altitude. Should there be only 2 more inches of manifold pressure of 200 RPM available (open the throttle to find out what's left), you can bet that the next fifteen minutes will find you in a negative performance situation. If you're flying at 8000 feet and the top is at 10, add whatever power is available, note the rate of climb that results (at normal climb airspeed), and figure out whether you can make it. Can do? Fine—have at it. Can't climb out of the ice?

Not fine, but make arrangements with ATC to go some-where else—NOW.

Even when you do escape the icing conditions, don't expect sublimation (the process by which ice "evapo-rates") to take place instantly. Sublimation time varies with outside air temperature, airspeed, and the amount and type of ice; this is one of the important reasons for carrying a larger-than-normal fuel reserve for winter-time IFR flights—there will be an inevitable need for more power to overcome the drag of structural ice.

An iced-up approach on instruments requires some special technique, too. Any ice accumulation other than a trace may cause strange aerodynamic things to happen, and unless you have a burning desire to test-fly an airplane of unknown characteristics, carry some ad-ditional speed throughout the approach, and be espe-cially wary of the circling maneuver. The extra air-speed may be enough to keep you honest while flying straight ahead, but a turn could be the aerodynamic straw that breaks your back.

Finally, "icing" is one of those magic words that usu-ally gets you anything you want from ATC; controllers recognize the severity of the situation and will nearly always grant any reasonable request for a different alti-tude. But suppose your request for climb or descent is denied, or put off until you know it will be too late? This is the time to exercise your authority. Tell ATC what you're *doing* (and why), and start on your way to a safe haven. Let *them* worry about traffic separation—that's their job.

HEAVY TURBULENCE

There will be a lot of IFR flights which turn out about two notches short of delightful because of turbulence —the kind of bumps and wallows which get certain stomachs churning, which require the near-constant attention of the pilot to keep the airplane where it belongs, but which needn't be considered anything more than mere annoyance. Our concern here is for the real McCoy, the heavy turbulence which can literally upset an airplane if the proper flight technique isn't applied. While turbulence of this magnitude is not infrequently associated with clear-air conditions (jet stream activity, mountain waves, and such), the damaging potential is heightened inside a thunderstorm, where the pilot has no choice but to rely on instrument displays to maintain some semblance of straight and level flight.

Most thunderstorm-related accidents are bad ones, with no one left to tell what really happened inside the storm; but there's not much doubt that when an airplane is upset "on instruments," the final damage is done when the pilot overstresses the airplane during a recovery attempt. So successful flight in heavy turbulence depends mostly on *not* getting upset.

At the top of the list, *numero uno*, is your determination to maintain the airplane in a level-flight attitude— or as nearly so as possible under the conditions at hand. Just like the highway-safety folks' admonition that "speed kills," there's a definite survival message about

61

airspeed: Too much and the airframe is likely to be overstressed and come unglued; too little and the wings will stall, with all the unhappiness that goes along with it. In between is a compromise known as *maneuvering speed,* a number which provides an angle of attack low enough to soak up sudden increases caused by upward gusts, yet high enough so that the wings will unload themselves when the point of overstress is approached. Happily, most light-aircraft maneuvering speeds (V_a for the technically minded) are not far from the high cruise airspeed determined in Chapter 1, so most pilots can rest assured that when the air gets lumpy enough to cause real concern, they're just about where they ought to be, airspeed-wise. (Check the Appendix where high cruise speeds are listed for a number of popular airplanes. If there's more than 20 knots or miles per hour difference between this and maneuvering speed, learn the pitch attitude and power setting which produce V_a, and use this combination in heavy turbulence.)

Once established, make no major changes in power or pitch; fly a constant attitude as if your life depended on it, which it probably does at this point. You will likely experience some awesome up- and down-drafts —thousands of feet per minute in a big storm cell—but resist the strong natural inclination to offset the resulting climbs and descents with pitch changes. Keep the little airplane on the attitude indicator as nearly level as possible, and when one of these big elevators in the sky takes hold of you, adjust power slightly to stop climbing or descending, and start a *trend* back to the original altitude. Don't worry about how much the altimeter winds around the dial; if there are other folks in there with you, they should be going up and down at the same rate, and you can only hope that vertical separation will be maintained. When you get a chance, let ATC know what's going on.

In summary, flying through really turbulent conditions should be avoided, but when it happens, maintain attitude as best you can, set the power for an airspeed close to maneuvering speed, and hang on. There's nothing more important than keeping the airplane upright, and the only way to do that is to fly attitude. Did you get the message? Here it comes again—average the attitude fluctuations, be satisfied with bracketing the indicated airspeed, don't make any big changes in anything, and FLY ATTITUDE!

VFR
APPLICATIONS

The Positive Flying philosophy fits hand-in-glove with the rather disciplined requirements of instrument operations—prescribed altitudes, headings, climbs and descents. But what about using these numbers in VFR conditions? How about those times when extra airspeed and an increased rate of descent will help you get home sooner, or put wheels to runway right on the spot you wanted? The three basic conditions of flight developed for instrument work need to be expanded and modified somewhat for visual operations, and that's the purpose of this chapter.

Let's take a look at the basic numbers as they apply

to a typical complex airplane, the Beechcraft Bonanza. This table shows the pitch attitudes, power settings, and airspeeds used for level flight, and climb or descent at 500 feet per minute:

PITCH ATTITUDE	MANIFOLD PRESSURE (2400 RPM)	AIRSPEED	VERTICAL SPEED
	20″	120 mph	500 fpm↑
	15″	120 mph	←→
	15″ + Gear	120 mph	500 fpm↓
	20″ + Gear	120 mph	←→
	20″	160 mph	←→
	15″	160 mph	500 fpm↓

With just two power settings, four pitch attitudes, two airspeeds, and judicious use of the landing gear, a Bonanza pilot can make his way through the most complicated instrument procedures. In addition, there are pitch and power settings to add 40 miles per hour for those who want to cruise and descend a little faster. Similar performance is possible with other airplanes; only the size of the numbers changes.

These basic IFR numbers can continue to be used very efficiently in visual conditions, when the pilot is much more at liberty to alter descent rates, airspeeds, and total performance to match the situation at hand. Again using the Bonanza as an example, here's the terminal portion of a typical VFR flight, starting with en-route cruise at 160 mph, manifold pressure at 20 inches, and the little airplane's nose right on the artificial horizon. When it's time to begin a 500 foot-per-minute descent (that rate is easy on *anybody's* ears), merely trim the pitch attitude to one nose-width below the horizon (this can be done very smoothly and positively with elevator trim alone), and leave the power setting unchanged; airspeed, which is now the variable, will increase to 180 miles per hour.

PITCH ATTITUDE	MANIFOLD PRESSURE	AIRSPEED	VERTICAL SPEED
	20″	180 mph	500 fpm↓

If that isn't a fast-enough descent, reduce power 5 inches more, for an additional 500 fpm with no retrimming—the attitude will drop another nose-width below the horizon to maintain the trimmed airspeed.

PITCH ATTITUDE	MANIFOLD PRESSURE	AIRSPEED	VERTICAL SPEED
	15″	180 mph	1000 fpm↓

There's a lot to be said for such a high-speed descent. It's obviously an efficient way to use the potential energy you've put into the airplane by climbing to a high cruise altitude (where it's cooler, clearer, and less congested), and the payoff shows up in decreased overall flying time. In addition to the time saved, the relatively high manifold pressure will keep the engine warmer

during the descent, and that's good for engine life—
your powerplant runs best when it runs hot, and aircraft engines are especially intolerant of rapid cooling.
Be *nice* to the engine.

Entering the terminal area (at least ten miles or so
from the airfield), it's prudent to slow down a bit, so
with careful use of elevator trim, raise the attitude to
one nose-width below the horizon, throttle still set at
15 inches. Airspeed will bleed off to approximately 160
mph, but the 500 fpm descent will continue.

PITCH ATTITUDE	MANIFOLD PRESSURE	AIRSPEED	VERTICAL SPEED
	15″	160 mph	500 fpm↓

At pattern altitude, decelerate to 120 miles per hour;
a slow increase in pitch attitude to one nose-width
above the horizon solves the problem *with no change
in power setting*—and no change in the very comfortable low-power noise level inside the cabin. (During this
transition, you're stopping the descent and decelerat-
ing, so raise the nose at a rate which will accomplish
these two objectives simultaneously.)

Somewhere on the downwind leg but no later than
a point opposite the intended touchdown (here's where
judgment comes in, and it can only be developed by a
lot of cutting and trying), lower the landing gear to
begin a 500 fpm descent—still at the same power setting. Should the situation require that you fly level on
downwind after the gear is lowered, raise the nose to

PITCH ATTITUDE	MANIFOLD PRESSURE	AIRSPEED	VERTICAL SPEED
	15″ + Gear	120 mph	500 fpm↓

one nose-width above the horizon and increase power to 20 inches.

PITCH ATTITUDE	MANIFOLD PRESSURE	AIRSPEED	VERTICAL SPEED
	20 ″ + Gear	120 mph	←→

Turning base, extend full flaps and retrim for an airspeed of 90–100 mph; the resultant sink rate should put you in the groove for a landing reasonably near the end of the runway. Does the approach look a little high? Reduce power for an increased rate of descent at the same airspeed. If by this time the engine is idling and you're still headed for the far end of the runway, push the nose down for 100 mph, or even 110—you'll come downhill much faster because of increased drag. On the other hand, an approach path which appears a bit on the low side (emphasis on "a bit") can be corrected by raising the nose just enough to drop the airspeed to 80 miles per hour, closer to best L/D airspeed and more efficient in terms of gliding distance. A *low* low approach should always be repaired by the addition of enough power to stop the descent until a normal glide path is once again intercepted; then reduce power by half the amount that was required to get you out of trouble. (By the way, if those last two procedures—increasing airspeed to correct a high approach, and slowing the airplane to stretch a short one—go against your aeronautical grain, stand by; you may be converted after you've finished the next chapter of *Positive Flying*.)

Now on final approach with full flaps, 90 miles per hour sets you up for a beautiful touchdown, but there are two things yet to be done; recheck the landing gear, and set the propeller control to maximum RPM as soon as airspeed and power are low enough to permit this without the prop winding up—noise abatement, you know. When the airplane is the proper distance above the runway (what's proper, you say?—the answer to that question is what makes aviation an *art*), combine a smooth increase in pitch attitude with a complete reduction of power and wait for that lovely squeaking noise.

A variation of this VFR procedure involves the situation in which you find yourself at 120 mph on downwind with the wheels hanging out, and the airplane ahead moseying along at 80. You can slow down easily —increase pitch attitude to three nose-widths above the horizon, set the power at 15 inches, and hang in there at 90 mph; when the spacing is more favorable and you're in position to finish your approach, extend the flaps and start down.

A well-planned VFR approach, one which considers wind, airport surface traffic, and distance from other airplanes in the pattern, should result in a completed

69

landing every time. But now and then it becomes necessary to abandon an approach, and the basic numbers come back into play. Any approach that is broken off before full flaps are extended should be considered a *missed approach* (for which you don't need 100 percent power), and the procedure is exactly the same as that used in IFR work: Increase power to 20 inches (or whatever climb-power number you've determined for your airplane), increase pitch attitude to two nose-widths above the horizon, and an immediate 500 foot-per-minute climb results.

A *go-around* is more urgent; it requires a maximum-performance climbout from a position close to the

CONDITION OF FLIGHT	PITCH ATTITUDE	POWER SETTING	AIRSPEED	VERTICAL SPEED
Level Cruise		20″	160 mph	⟷
Begin Descent (Retrim)		20″	180 mph	500 fpm↓
Come Down Faster		15″	180 mph	1000 fpm↓
Enter Terminal Area (Retrim)		15″	160 mph	500 fpm↓
Enter Downwind (Retrim)		15″ + Gear	120 mph	500 fpm↓
Level on Downwind		20″ + Gear	120 mph	⟷
Base Leg		15″ + Gear + Flaps	90 mph	⟷

Final Approach: Pitch, Power, Airspeed as Required to Produce and Maintain the Desired Glide Path.

shined a light into the dim corners of their airplane's performance envelope are likely to find some unpleasant surprises there. Accidents related to short fields and last-minute go-arounds seem to confirm that pilots try to get more performance from their airplanes than was built in, or fail in the pilot-technique department, or both. As in no other aeronautical arena, maximum-performance situations have got to be handled right the first time.

The backbone of Positive Flying—attitude plus power equals predictable performance—prevails throughout this chapter, perhaps even more significantly than in the others. In addition to developing very precise attitudes and configurations for maximum performance, we are confident that at least two other good things will happen: You'll be a more precise pilot by the time you've run through all the drills in this chapter, and a comfortable aura of confidence—in your airplane and yourself—will surround all your flight operations from now on. There's a solid, genuine, pilot-in-*command* feeling when you know exactly what the airplane will do and how to go about making it happen.

EXPANDING THE BASIC NUMBERS

No matter what kind of airplane you're flying, some convenient and easy-to-remember relationships among airspeed, power setting, landing gear, and vertical speed should have become apparent by now. Using the Bonanza at 120 mph as an example, a power change of 5 inches produces either a 40 mph increase in airspeed or a 500 fpm change in vertical speed; retract (or lower) the landing gear, and the airplane will climb (or descend) at 500 fpm. (The convenience and reliability of these units of change is one reason that 120 was chosen as the working speed for the Bonanza —and was a major consideration in determining the speeds recommended in the Appendix for other airplanes.)

If the landing gear or a 5 inch power change is worth 500 fpm or 40 mph, it follows that a power increase to 25 inches in level flight should result in the airplane accelerating to 200 mph; and on the low-speed end, extending the landing gear should drop the airspeed from 120 to 80. But such is not the case, because at speeds much above or below 120, total drag begins to change significantly. The Bonanza pilot will find that the top-end speed with an additional 5 inches of power is about 180; if he lowers the wheels in level flight at 120, airspeed will not suffer by 40 mph, but will settle down at 90. There is a noticeable difference in drag in both instances, confirming the aerodynamic truth that drag changes as the *square* of the velocity—double the airspeed and you *quadruple* the drag; cut airspeed in half, and drag is only one-fourth what it was.

We are not particularly interested in the drag behavior of the airplane at speeds above normal cruise, but if the wheel-drag keeps getting less and less at lower speeds, can we reach a point where the landing gear produces so little drag that there's *no* change in airspeed? The answer is undoubtedly yes, but there's a much more important quality to be investigated; at a given airspeed, what effect does landing gear have on *vertical speed?* Can the minimal drag of low speeds be translated into more *climb?* You bet it can; so it's back to the drawing board to find out how much. (Pilots of fixed-gear airplanes, read on—the lesson is worth learning. Besides, you may move up to a retractable in the future, when you can put this knowledge to good use.)

With a clerical assistant in the right seat, trim the airplane carefully for its normal working airspeed— "low cruise" from the Appendix—and lower the landing gear. Record the rate of climb or descent as soon as vertical speed is stable (it doesn't matter whether you're going up or down when you start). Some aircraft with complex gear-door arrangements (Cessna's Skymaster is a *beaut!*) tend to oscillate somewhat as the doors are operating, and you may have to help by damping the pitch excursions; the vertical speed reading should not be recorded until the airplane is climbing or descending at a steady airspeed—the one for which it was trimmed.

Now reduce airspeed in 10 mph increments with the gear extended, and note the *change* in vertical speed with each reduction. When the airspeed approaches the bottom of the white arc, extend enough flaps to

75

lower the stall speed to a safe point, and continue the experiment. The Bonanza produced this table of airspeed and vertical speed changes:

AIRSPEED (MPH)	CHANGE IN VERTICAL SPEED (FPM)
60	0
70	50
80	100
90	200
100	300
110	400
120	500

If nothing else, this exercise should prove rather conclusively that there's precious little to be gained by pulling up the wheels at very low airspeed during a go-around; there are more important things to be doing at a time like that anyway. And if you should suffer the indignity of actually touching down when you really meant to keep on flying, less damage will be done (to your passengers, the airplane, and your pride) if such an inadvertent landing occurs with the wheels extended.

While we're here, there are some equally interesting high-speed considerations of landing-gear effect. A Bonanza with its wheels down will descend at 500 fpm when trimmed for 120 miles per hour; but increase the speed to 140, and the V-tailed wonder heads for the ground at 1000 fpm. At 160 mph, the rate of descent will increase to 1500 fpm, and if the throttle is reduced all the way to idle—20 inches of manifold pressure maintained up to this point—the additional 2000 foot-per-minute descent produces an astounding earthward velocity of *3500* fpm! So when you need to descend in a hurry, get the gear down, close the throttle, and push the nose over until gear-limit speed is reached. If you need still more vertical speed, set the prop control to high RPM, put your hands out the window, and think *heavy.*

DON'T LET YOUR RECORDING
SECRETARY GO HOME YET. . .

. . . because in the next exercise, perhaps the most important one in the series, precise airspeed control is a must; you should have a second pair of eyes for the vertical speed readings. (Never let it be said that Positive Flying is not a social event.)

Having just proved that there is a speed at which the drag of the landing gear has absolutely no effect on vertical speed, let's dig a little deeper into your airplane's performance capabilities and find out about its actual performance—not just the changes—in the "full dirty," low-speed configuration. The most meaningful information will be that obtained with the airplane at its maximum allowable gross weight, or at least the heaviest load you normally carry. Although the gear-drag demonstration is not affected by weight, the actual performance will vary remarkably as pounds are added or taken away. Use sandbags or people, but load the airplane realistically—anything less will provide a set of dangerously misleading numbers.

For safety's sake, start this drill several thousand feet above the terrain. Note the temperature at level-off, because the computation of performance altitude (density altitude, if you prefer) will be just as important as the numbers which result.

All set now? Lower the landing gear, extend full flaps, and let the airspeed bleed off (it won't take long!) to the lowest number used in the gear-drag experiment. At this point, advance the throttle to full power —the mixture may need to be adjusted—and while you very carefully maintain the entry airspeed, have your observer record the vertical speed as soon as it settles

RATE OF CLIMB WITH FULL POWER, LANDING GEAR, AND FULL FLAPS (PERFORMANCE ALTITUDE, 4000 FEET)	
AIRSPEED (MPH)	OBSERVED RATE OF CLIMB (FPM)
60	400
70	500
80	500
90	400
100	200
110	0
120	−300
130	−600

down. When the reading has been taken, lower the nose smoothly (still at full power) until airspeed increases exactly 10 mph and take another reading. Continue the airspeed increases in 10 mph increments until you've reached the maximum flap-extension speed, then clean up the machine and take a look at the figures you've obtained. Our Bonanza turned out like in the chart above.

Those numbers should get your attention. They should leap off the paper at you with the message that the airplane will perform positively—*i.e.*, will *climb*—in the landing configuration at an airspeed much lower than you would ever have dreamed of using for a go-around. When the situation is grim and you must climb away, airspeed is your salvation, right? Right—as long as the airspeed you use is a *very low one;* and now you know what that number is. If your airplane displayed even a small amount of climb at some point on the airspeed scale you've just developed, you have proved that positive go-around performance can be obtained in the worst possible situation. The strong urge to go for more airspeed (it's deeply entrenched—we've been taught from the beginning that flying at these super-slow speeds is tantamount to disaster, especially at heavy weights) can in fact be your undoing; total drag builds up and prevents the airplane from climbing as well as it might, and obviously, at the higher speeds, drag guarantees a *descent*—something you don't need in the middle of a go-around.

For comparison, and thinking of the possibility of a

go-around at a mountain airport on a hot day, we took the Bonanza to a performance altitude of 10,000 feet and went through the exercise again—same airspeeds, same weight. Here's what showed up:

RATE OF CLIMB WITH FULL POWER, LANDING GEAR, AND FULL FLAPS (PERFORMANCE ALTITUDE, 10,000 FEET)	
AIRSPEED (MPH)	OBSERVED RATE OF CLIMB (FPM)
60	100
70	200
80	200
90	100
100	−100
110	−300
120	−600
130	−900

Not exactly a skyrocket, but it's nice to know that the airplane *will climb*—albeit less spectacularly—even under such drastic conditions. It's very simple: Attitude plus power equals predictable performance.

The go-around demonstration can be visualized more easily by putting the numbers on a graph. With rate of climb on the vertical axis, and indicated airspeed across the top, the 4000- and 10,000-foot curves came out as shown on the graph on page 81. Notice that this airplane exhibits the same climb capability—or level flight, if power were reduced a bit—at two airspeeds: 60 and 90 mph, for example. That's an indication of the much-argued aerodynamic "step," and

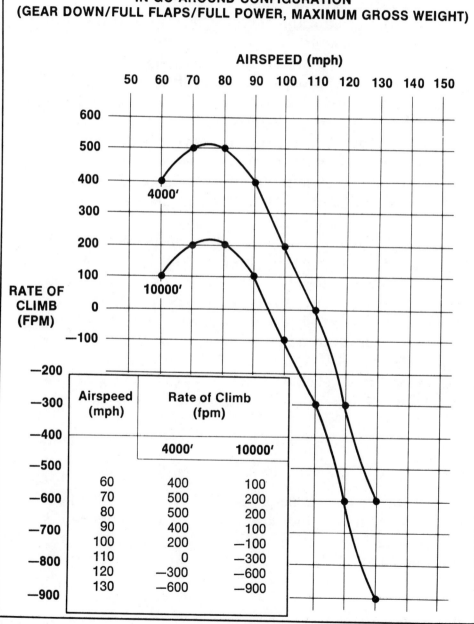

CLIMB PERFORMANCE AT VARIOUS AIRSPEEDS & DENSITY ALTITUDES IN GO-AROUND CONFIGURATION
(GEAR DOWN/FULL FLAPS/FULL POWER, MAXIMUM GROSS WEIGHT)

Airspeed (mph)	Rate of Climb (fpm)	
	4000'	10000'
60	400	100
70	500	200
80	500	200
90	400	100
100	200	−100
110	0	−300
120	−300	−600
130	−600	−900

when an airplane is so heavily loaded or under-engined that full power is required to sustain level flight, the step becomes pretty important.

At the peak of the curves is the airspeed that will produce maximum performance in the full-flap, gear-extended, full-power configuration; it's the point at which the compromise between lift and drag is most favorable with a "dirty" airplane, and it is no doubt very close to the go-around speed quoted in the operating handbook. The airspeed is obviously the same for the two performance altitudes chosen, and it will be the same for *all* altitudes; the difference, of course, is the amount of performance (in this case, rate of climb) that will be produced.

You have now determined an airspeed and pitch attitude guaranteed to produce the best go-around performance possible. You have also proved that there's no need to worry about which to retract first —gear or flaps—because full power and a predetermined pitch attitude will do the trick *even with gear and flaps fully extended.* The Positive Flying go-around is as reliable at Leadville, Colorado, as at the Dead Sea—power plus attitude always equals predictable performance.

The go-around pitch attitude is extremely important; you should burn that pitch attitude picture into your mind. You should also make a note of it, and add that picture to your table of attitude/power combinations. If your attitude indicator is marked with lines representing various pitch angles in degrees, so much the better. Regardless of the attitude display, don't hesitate when a go-around becomes necessary; remember, we're not talking about missed approaches but bona fide, right now, if-we-don't-start-climbing-we're-gonna-hit-something go-arounds! Increase power to FULL, raise the nose to that ever-faithful pitch attitude, and watch your performance prediction come true. You may have to adjust pitch slightly to get the precise airspeed you need, but the predetermined attitude is a good place to begin.

The go-around pitch attitude will be a high one—it will remind you of the attitude just prior to a power-on stall—but hang the nose up there where your tests proved it needs to be, and *always cross-check the airspeed indicator for the proper reading;* at this point, it's your key to success.

If the demonstrated maximum go-around performance of your airplane turns out to be something less than what's needed, you should give serious thought to two possible solutions; one involves some inconvenience, the other a lot of dollars. First, you can always reduce the load—fewer people, less fuel or baggage—to the point where there's enough performance to overcome the obstacles in your path. The second consideration (the expensive one) concerns power available; the incontrovertible fact is that a normally aspirated engine loses power as altitude increases. If you are flying behind such an engine and it just doesn't cut the mustard in the thin, high-altitude air you frequent, there's little choice—you need turbos to get the job done.

83

KEEP A GOOD THING GOING

A go-around is only as successful as the climbout which follows, so you must "clean up" the airplane and allow it to accelerate to normal climb speed, where the wing can do its best work. We said earlier that there was no need to *worry* about whether to pull in the gear or flaps first, and there is likewise no need to *hurry* once you've achieved a safe height above the terrain. At go-around airspeed there is precious little to be gained drag-wise by retracting either, but as speed builds, the benefits of retraction begin to show up; check your gear-drag table to prove the point.

At full extension, wing flaps produce far more drag than lift, suggesting that they should come up first, followed shortly thereafter by the landing gear.* You must make adjustments in pitch attitude as the flaps retract, but you *will* be able to maintain a positive rate of climb throughout. (If the airplane is one of those for which partial flap extension is recommended for normal takeoff and climb, be certain to stop the flaps when that position is reached.)

And be ye not concerned about stalling, if the flaps are "milked" up—that is, raised in small increments instead of all at once. By allowing airspeed to build a bit between retractions, you'll soon be well above the

---

*As a general rule, the flap extension recommended by the manufacturer for a short-field takeoff is the setting at which the most favorable lift-drag ratio exists.

speed at which the airplane will stall in the clean configuration.

Of course, it becomes necessary to know just what that stalling speed is, which leads into the next exercise in this series of exploring your airplane's maximum-performance capabilities. Starting at a safe altitude (3000–4000 feet above the ground), and with your observer's pencil at the ready, go through a complete straight-ahead stall series in power-on and power-off configurations. You need two precise readings for each situation: the airspeed at which the stall warning activates, and the airspeed at which the stall actually occurs. Don't cheat; wait until a *full* stall happens. If it's been a long time since you actually stalled an airplane with full power, you might consider hiring a qualified instructor to go with you; this makes for an expensive note-taker, but it's a small price to pay for the additional safety factor. Our Bonanza stall warning/stall speed experiment came out like this:

	HORN SOUNDED	STALL OCCURRED
POWER OFF (Idle)		
Clean	80 mph	70 mph
Gear + Full Flaps	70 mph	60 mph
FULL POWER (Properly Leaned)		
Clean	75 mph	65 mph
Gear + Full Flaps	65 mph	50 mph

Right now, the most important number is the one in the lower right-hand corner of the table, the speed at which the airplane quit flying in the gear-extended, full-flap configuration. It's important because it confirms the margin between stall speed and go-around speed. And in light of the flap-retraction discussion just concluded, notice the *clean* stall speed—there's plenty of room between that number and the airspeed you'll be looking at by the time flap retraction is completed.

BUT THERE'S A MOUNTAIN AHEAD!

Nearly all the situations which require a true go-around (as defined earlier) can be placed in two categories: First, about three microseconds before tires meet concrete in what was going to be the sweetest landing you've ever made, the controller hollers "Go-around!" because somebody else is trying to use your runway at the same time; second, your approach turns out to be a little long, the runway a little short, and you realize that discretion—otherwise known as a go-around—is going to be the much better part of valor today.

Common to each of these situations is the need to climb away from the ground and turn away from whatever lies ahead. So there you are, confident that the Positive Flying go-around will solve the *climb* part of your problem, but unsure—if not terrified—about the prospect of rolling the airplane into a bank steep enough to clear the obstacle under the nose.

86

There's no doubt that stall speed increases as the airplane is banked, but rather than proceed into this "critical" situation with only a guesstimate, perform the following exercise and find out exactly what the figures will be for your airplane. The situation may not be as critical as you have been led to believe. In addition, this exercise will help you understand *why* the stall speed gets higher in a turn.

This is a hands-off demonstration, so the first order of business is complete and proper elevator trimming; set the airplane up in level flight at an airspeed approximately 10 mph above the clean, power-off stall warning speed—we used 80 mph in the Bonanza. Now, ever so slowly and as smoothly as you are able (it's best to do this on a day when there's not so much as a ripple in the air), roll the airplane into a 15-degree bank with rudder pressure alone—there must be absolutely no control-column inputs.

With the very first degree of bank, a small portion of the wing's total lifting force is diverted to the job of pulling the airplane around the turn, and as bank increases, this portion grows accordingly. Elevator trim, set up for 80 mph in *wings-level flight,* will lower the nose until airspeed increases just enough to replace the lift that was lost to the turning force; in the Bonanza, 15 degrees of bank required an additional 5 miles per hour. Increase the bank to 30 degrees, and airspeed slides up another 5 mph.

At 45 degrees of bank, the logarithmic increase in airspeed starts to show; instead of the 5 mph changes experienced up to this point, the airspeed will *double* its increase for the additional 15 degrees of bank, and the same thing will happen if you continue to a 60-degree bank; you'll also feel a seat pressure of 2 Gs.

The object here is not to explore your personal G-force tolerance, but to illustrate that, in effect, the airplane becomes heavier in a turn, and that the stall speed goes up right along with this higher effective weight. How much? The same amount as the increase in trim airspeed. In other words, when your airplane was banked 15 degrees and increased its speed 5 miles per hour to make up for the loss of vertical lift, stall speed also increased by 5 mph because of the bank; at 30 degrees, stall speed went up another 5 mph, and so on. It stands to reason that stall speed will show similar increases regardless of aircraft configuration, so now you can compare the go-around situation at several angles of bank and determine just how much you can bank to miss the mountain and still be safely above a stall. Here are typical Bonanza numbers:

TRIM AIRSPEED	BANK ANGLE	STALL SPEED (In Go-around Configuration: Gear Down, Full Power, Full Flaps)
MPH	DEGREES	MPH
80	0	50
85	15	55
90	30	60
100	45	70
120	60	90

Now you have a set of numbers which can be applied to what was learned earlier—that the best airspeed for a go-around is 77 miles per hour (in a fully loaded Bonanza).

BANK ANGLE	STALL SPEED (In Go-around Configuration: Gear Down, Full Power, Full Flaps)	SAFETY MARGIN (Difference Between Go-around Airspeed and Stall Speed)
DEGREES	MPH	MPH
0	50	27
15	55	22
30	60	17
45	70	7
60	90	TILT!

It's obvious from this comparison that the maximum bank angle to guarantee a comfortable airspeed pad is 30 degrees—and you can do a lot of turning at 77 mph and 30 degrees of bank. The radius of such a turn will be about 2600 feet (in *any* flying machine at this speed), which means that you can do a 180 degree turn and get the heck out of trouble in half the length of a mile-long runway.

TAKEOFFS FROM
SHORT AND/OR SOFT FIELDS

Each pilot has his own definition of the "short field"; like emergencies, near-misses, and the bar-side recounting of heroic adventures in the air, it is a very subjective thing. In addition to perceived runway length under a specific set of conditions (width, proximity, lighting, slope, surrounding terrain, etc.), decreased air density effectively shrinks the number of feet available for getting a flying machine into the air—the field that was long enough this morning, or last week, or last winter, may be completely inadequate today. Whatever the reason, this discussion of techniques for short- and soft-field operations deals with *any* situation in which supernormal performance is required to accomplish a successful takeoff.

Your first consideration must be the actual length of the takeoff surface (published figures are usually reliable, but step it off if there's any doubt—and don't fudge on the size of your strides or the value assigned to each step!) compared to the minimum-run distance quoted in the operating handbook. When the distance required is more than, or very close to, the distance available, the most effective—by *far* the most effective—procedure is to lighten the load; full power can only lift so much. Two trips, while perhaps inconvenient, are a hell of a lot better than half a takeoff.

With the airplane properly loaded, don't close the handbook until you look up the recommended flap setting for a short-field takeoff. Whatever the manufacturer quotes is the proven best compromise of lift and drag generated by the wing flaps; using a ball park estimate will surely extract its price, one way or the other.

Now the stage is set. The airplane weighs no more than it can safely carry from the runway available, the flaps are positioned to produce the most lift at the lowest possible airspeed, you have confidence that the airplane can do what needs to be done, and all that remains is the application of full power—which is every bit as important as the other factors in a short-field takeoff. At performance (density) altitudes of 2000 feet or more, there will be a significant and unavoidable loss of power from nonturbocharged engines, and it becomes vital to lean the mixture if you are to achieve the performance figures quoted in the operating handbook. Hold the brakes, run the engine to full throttle, and lean for maximum power. (For fixed-pitch installations, lean to maximum RPM; for constant-speed props, adjust the mixture control until the fuel-flow or fuel-pressure indicator shows the proper number. In any case, check the operating handbook for the correct procedure.)

Count the final bead, take a deep breath, and release the brakes; when you can hold the nose straight with rudder alone, rotate to the same pitch attitude you used for go-around. (This will require some practice, preferably on a very wide runway which is long enough to tolerate your directional-control education. Remember that nosewheel steering will be lost early in the takeoff rolls, and if at that point there's not enough airflow to keep things straight with the rudder, you're going to head for the boondocks. Wind direction and velocity make a big difference here, too; a right crosswind tends to help, while a wind blowing from the left will force you to keep the nosewheel on the ground for positive steering until a higher airspeed is reached. If it takes *full* rudder, use it.) At this very low airspeed, it's likely that full-up elevator will be needed to get the nose to the proper attitude, but be very much aware that as the airplane accelerates, this big elevator displacement will become more and more effective. When the pitch attitude looks right, keep it there with whatever control inputs are required.

At liftoff (the earliness of which may surprise you, especially with a decent wind blowing down the runway), don't worry about pulling up the wheels or retracting the flaps; you proved earlier that neither of them produce significant drag at go-around airspeed. Keep the airplane flying straight ahead (or turning sufficiently to miss obstacles—30 degrees of bank still looks good as a maximum), and maintain the airspeed that your experiments showed to be the best for maximum rate of climb. When all the nasty things are below and behind, ease the nose down a bit for acceleration toward best-climb airspeed, and retract flaps and gear at your discretion.

Surely the most dramatic proof of this short-field pudding came as the result of a long-distance telephone call a few years ago. Having wisely landed in a corn field when trapped between cloud-shrouded Mexican mountains and the onset of darkness, a pilot rang up his instructor at 2 A.M. New York time ("You're calling from *where?*"), explained the situation, and wondered if he should consider flying the airplane out. After the length of the field—900 feet—and its elevation above sea level—6500 feet—were determined, and having established that the owner-pilot did not want to reimburse *los Mexicanos* for taking off the wings and carting the airplane into town, the instructor proceeded to instruct in a most unique fashion: He explained the entire Positive Flying short-field takeoff procedure—proper pitch attitude, proper airspeed, full power—to his long-distance student, courtesy of Ma Bell and her southern *compadres*. With just enough fuel in the tanks to get to a real airport, everything loose (including flashlight and log books) removed from the airplane, and the passengers sent to Mexico City by bus, the pilot executed a successful super-short-field takeoff. *Positive Flying works.*

IT RAINED HARD ALL NIGHT

The Positive Flying soft-field technique takes into consideration only the condition of the takeoff surface —it could be soft, rough, unpaved, or all three—and the objective is to transfer the weight of the airplane from wheels to wings as soon as possible. A soft, *short* field presents an even greater challenge to your piloting skills; the techniques may need to be combined, and you must keep in mind that long grass, deep snow, or soft dirt/sand/gravel will probably increase the takeoff distance by 100 percent—or more.

A soft-field takeoff begins with the configuration used

for getting out of a short field (same flap setting, full power), the only difference being that you should bring the nose to the go-around attitude *earlier* and let the airplane fly off when it's ready. (The early high attitude also gets the nosewheel out of the soft stuff sooner, reducing drag remarkably.) Since you don't know the exact stall speed, you are merely providing an angle of attack which will allow the wings to fly when sufficient airspeed is generated, and when that occurs, *voilà!*— aviation at the lowest possible speed. If the go-around attitude is maintained religiously throughout, there is no danger of stalling when you leave the ground; the airplane won't fly unless and until an adequate angle of attack is reached, and that angle will always be reduced when liftoff occurs. A safe margin above stall is therefore provided automatically *if* you are careful to maintain the go-around attitude and airspeed.

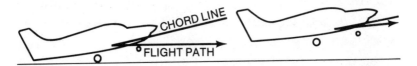

Angle of attack is reduced at liftoff because of the change in flight path.

LANDING ON
SHORT AND/OR SOFT FIELDS

Perhaps the most overrated of the maximum-performance maneuvers, a short-field landing simply adds rapid on-the-ground deceleration to a normal full-flap, full-stall landing. True, there may be approach-end obstacles which require a steep glide path, forcing you to waste some of the landing surface, but in the final analysis, the success of bringing your airplane to a stop within the space available depends mostly on the point of touchdown. It's precise flying, and it's what spot landings are all about; if you can bring the spot and the stall together at the same time with enough distance remaining for the brakes to stop the airplane, you've won the short-field battle.

Good judgment and piloting technique are heavily involved in short-field landings, and the only way to develop those qualities is practice. Since you know how your airplane handles in the full-flap, low-airspeed, maximum-performance configuration, start with approaches to a long runway and with an airspeed on final somewhat higher than the go-around number—add 10 miles per hour for openers. When you can touch down on the same spot each time, begin slowing your approaches until you are flying at the go-around airspeed; use runway lights or whatever's available to note the actual distance required to stop the airplane, and after a few tries you should be pleasantly surprised to find that your short-field performance can be really *short!*

As soon as the main wheels make contact with the ground, hold the control column all the way back—the airplane should be fully stalled by now—to keep as much weight as possible on the tires for maximum braking power, and to develop as much form drag as possible. You'll soon discover that most light aircraft, when landed full-flap, full-stall, and nose-high, will require very little stopping distance, even *without* brakes; a field short enough to require maximum braking may be a field that's too short for the airplane. It's a common and expensive trap—you can often get into a field a lot easier than you can get out.

The approach to a soft landing surface should be no different than the short-field technique, but hold off on the brakes after touchdown. The primary consideration is to touch down at the lowest possible ground-speed, then hold the nosewheel off until there's no more elevator power. The nosewheel is not only the weakest link in the undercarriage structure, it also becomes a very effective pivot point for the entire airplane; if the nosewheel digs in while there's still enough speed remaining, you may suddenly find yourself looking back down the runway from the inverted position.

Practicing the skills necessary for short-field operations is one thing; being faced with the real situation is another. It's amazing how trees and telephone poles reach upward, how that grass strip shrinks before your eyes as you approach for the first time. Most bush pilots make a reconnoitering pass before committing themselves, even at familiar landing sites; if it's good enough for them, a practice run ought to be considered by anyone with less experience and know-how. And when you're right down to the brass tacks and things just don't look right, *go around!* You are all set up to do exactly what you have practiced, and you know that full power will start an immediate climb. When in doubt, do it!

BEFORE
YOU GET INTO
THE APPENDIX . . .

We began *Positive Flying* with a discussion of "flying by the numbers," and have made repeated reference to the fact that when an airplane is placed in a certain pitch attitude and a certain amount of power is applied, the performance of that airplane is predictable—and in the Appendix which follows, those attitudes, power settings, and the resulting performance are documented for the most popular general aviation aircraft. For each model, you'll find the numbers for basic instrument procedures, plus a high-speed descent and the go-around. We feel very strongly that pilots should be aware of *why* their airplanes will perform positively in the go-around configuration, and so we've included the tables and graphs which show the derivation of those numbers.

If you test yourself and your airplane against the information in the Appendix, please bear in mind that these are to be considered *starting* figures—targets, if you will. They were developed using aircraft which ranged from brand new to used-hard-and-put-away-wet, and in some instances we were operating at slightly less than maximum gross weight; therefore, the numbers produced by your airplane may differ somewhat. The major factors to be recognized and allowed for are engine condition, density altitude, and gross weight, so please take these into consideration when checking your numbers against ours.

Certain models exhibit a significant difference between indicated and calibrated airspeeds; this will show up most remarkably at the lower end of the airspeed scale, where the use of the wrong number is most critical. Because all of the performance figures herein are based on indicated airspeed, we suggest that as you put your airplane through the Positive Flying paces, you find out *first* exactly where the stall occurs. Once determined, the proper stall speed (indicated) provides a base for the rest of the numbers.

(We're aware of one old-timer who regularly flew a single-engine retractable from very short fields, getting outstanding performance from his flying machine. When asked what airspeeds he used, he claimed that he never saw anything less than 100 mph on the indicator during a short-field approach. A bit of research showed that his airspeed indicator was just about 20 miles per hour wrong at approach speeds; the old boy had been flying it right, but his numbers were a long way from what the book—*any* book—recommended.)

At the end of the Appendix are a couple of blank forms for those readers who are operating airplanes not included in this list—new models, modifications, home-builts, and so on. By following the procedures outlined in Chapter 2, "Nailing Down the Numbers," you can come up with an Appendix page of your own—even if you fly a Sopwith Camel. Please, be our guest.

APPENDIX
The Positive Flying
Numbers and Performance Figures for
27 Popular General Aviation Aircraft

TABLE OF CONTENTS

BEECHCRAFT MUSKETEER   150 HP

ATTITUDE	POWER		PERFORMANCE	
	MAP/Landing Gear	RPM	Airspeed (mph)	Vertical Speed (fpm)
THE NUMBERS FOR BASIC IFR				
		2600	90	500 ↑
		2300	90	0
		2000	90	500 ↓
		2600	110	0
		2300	110	500 ↓

ATTITUDE	POWER		PERFORMANCE	
	MAP/Landing Gear	RPM	Airspeed (mph)	Vertical Speed (fpm)

THE NUMBERS FOR HIGH-SPEED DESCENT

ATTITUDE	POWER		PERFORMANCE	
		2600	120	500↓
		2300	120	1000↓

BEECHCRAFT MUSKETEER   150 HP

4

MAXIMUM SAFE BANK ANGLE DURING GO-AROUND	
Angle of Bank	**Safety Margin***
0°	16 mph
15°	11 mph
30°	6 mph
45°	−4 mph

* This represents the difference between stall speed with gear down/full flaps/full power, and the optimum go-around airspeed of 60 mph.

5

THE NUMBERS FOR MAXIMUM PERFORMANCE (GO-AROUND)				
ATTITUDE	**POWER**		**PERFORMANCE**	
SEA LEVEL	**MAP/Landing Gear**	**RPM**	**Airspeed (mph)**	**Vertical Speed**
8000'	FULL POWER FULL FLAPS		60	MAXIMUM

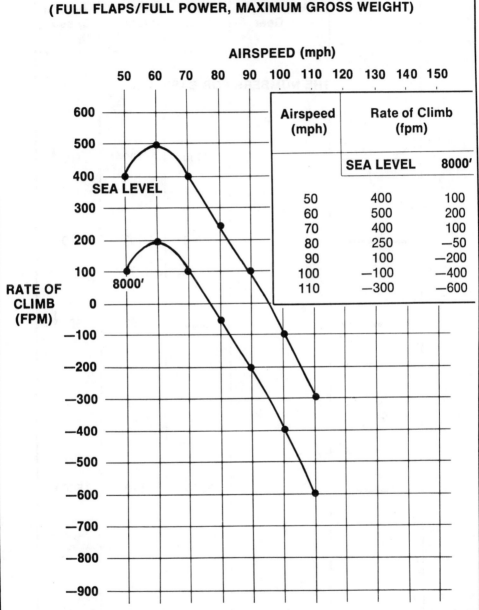

6

CLIMB PERFORMANCE AT VARIOUS AIRSPEEDS & DENSITY ALTITUDES IN GO-AROUND CONFIGURATION
(FULL FLAPS/FULL POWER, MAXIMUM GROSS WEIGHT)

Airspeed (mph)	Rate of Climb (fpm)	
	SEA LEVEL	8000'
50	400	100
60	500	200
70	400	100
80	250	−50
90	100	−200
100	−100	−400
110	−300	−600

107

BEECHCRAFT SUNDOWNER   180 HP

1

ATTITUDE	POWER		PERFORMANCE	
	MAP/Landing Gear	RPM	Airspeed (mph)	Vertical Speed (fpm)
THE NUMBERS FOR BASIC IFR				
		2600	100	500 ↑
		2300	100	0
		2000	100	500 ↓
		2600	120	0
		2300	120	500 ↓

2

ATTITUDE	POWER		PERFORMANCE	
	MAP/Landing Gear	RPM	Airspeed (mph)	Vertical Speed (fpm)
THE NUMBERS FOR HIGH-SPEED DESCENT				
		2600	130	500 ↓
		2300	130	1000 ↓

109

BEECHCRAFT SUNDOWNER   180 HP

4

MAXIMUM SAFE BANK ANGLE DURING GO-AROUND

Angle of Bank	Safety Margin*
0°	25 mph
15°	20 mph
30°	15 mph
45°	5 mph

* This represents the difference between stall speed with gear down/full flaps/full power, and the optimum go-around airspeed of 75 mph.

5

THE NUMBERS FOR MAXIMUM PERFORMANCE (GO-AROUND)

ATTITUDE	POWER		PERFORMANCE	
	MAP/Landing Gear	RPM	Airspeed (mph)	Vertical Speed
SEA LEVEL				
8000'	FULL POWER FULL FLAPS		75	MAXIMUM

110

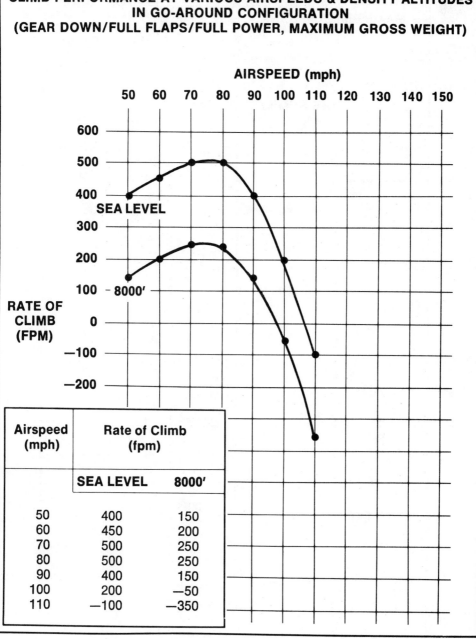

CLIMB PERFORMANCE AT VARIOUS AIRSPEEDS & DENSITY ALTITUDES IN GO-AROUND CONFIGURATION (GEAR DOWN/FULL FLAPS/FULL POWER, MAXIMUM GROSS WEIGHT)

Airspeed (mph)	Rate of Climb (fpm)	
	SEA LEVEL	8000'
50	400	150
60	450	200
70	500	250
80	500	250
90	400	150
100	200	−50
110	−100	−350

1

ATTITUDE	POWER		PERFORMANCE	
	MAP/Landing Gear	RPM	Airspeed (mph)	Vertical Speed (fpm)

THE NUMBERS FOR BASIC IFR

ATTITUDE	MAP/Landing Gear	RPM	Airspeed (mph)	Vertical Speed (fpm)
		2300	100	500 ↑
		2000	100	0
		1700	100	500 ↓
		2400	130	0
		2100	130	500 ↓

112

2

ATTITUDE	POWER		PERFORMANCE	
	MAP/Landing Gear	RPM	Airspeed (mph)	Vertical Speed (fpm)

THE NUMBERS FOR HIGH-SPEED DESCENT

ATTITUDE	POWER		PERFORMANCE	
		2400	145	500 ↓
		2100	145	1000 ↓

4

MAXIMUM SAFE BANK ANGLE DURING GO-AROUND

Angle of Bank	Safety Margin*
0°	17 mph
15°	12 mph
30°	7 mph
45°	—3 mph

* This represents the difference between stall speed with gear down/full flaps/full power, and the optimum go-around airspeed of 75 mph.

5

THE NUMBERS FOR MAXIMUM PERFORMANCE (GO-AROUND)

ATTITUDE	POWER		PERFORMANCE	
	MAP/Landing Gear	RPM	Airspeed (mph)	Vertical Speed
SEA LEVEL				
8000'	FULL POWER FULL FLAPS		75	MAXIMUM

CLIMB PERFORMANCE AT VARIOUS AIRSPEEDS & DENSITY ALTITUDES IN GO-AROUND CONFIGURATION
(FULL FLAPS/FULL POWER, MAXIMUM GROSS WEIGHT)

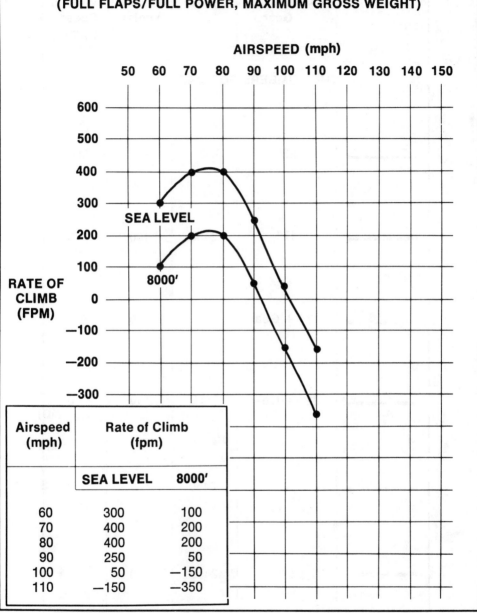

Airspeed (mph)	Rate of Climb (fpm)	
	SEA LEVEL	8000'
60	300	100
70	400	200
80	400	200
90	250	50
100	50	−150
110	−150	−350

BEECHCRAFT SIERRA   200 HP

1

ATTITUDE	POWER		PERFORMANCE	
	MAP/Landing Gear	RPM	Airspeed (knots)	Vertical Speed (fpm)

THE NUMBERS FOR BASIC IFR

ATTITUDE	MAP/Landing Gear	RPM	Airspeed (knots)	Vertical Speed (fpm)
	22"	2450	100	500 ↑
	17"	2450	100	0
	17" + Gear	2450	100	500 ↓
	22" + Gear	2450	100	0
	22"	2450	120	0
	17"	2450	120	500 ↓

ATTITUDE	POWER		PERFORMANCE		**2**
	MAP/Landing Gear	RPM	Airspeed (knots)	Vertical Speed (fpm)	

THE NUMBERS FOR HIGH-SPEED DESCENT

ATTITUDE	MAP/Landing Gear	RPM	Airspeed (knots)	Vertical Speed (fpm)
⊂⊐	22″	2450	135	500 ↓
⊂⊐	17″	2450	135	1000 ↓

3

LANDING GEAR DRAG AT VARIOUS AIRSPEEDS

Airspeed (knots)	Change in Vertical Speed (fpm)
60	0
70	100
80	300
90	400
100	500

117

4

MAXIMUM SAFE BANK ANGLE DURING GO-AROUND	
Angle of Bank	**Safety Margin***
0°	17 knots
15°	12 knots
30°	7 knots
45°	—3 knots

* This represents the difference between stall speed with gear down/full flaps/full power, and the optimum go-around airspeed of **65** knots.

5

THE NUMBERS FOR MAXIMUM PERFORMANCE (GO-AROUND)				
ATTITUDE	**POWER**		**PERFORMANCE**	
	MAP/Landing Gear	**RPM**	**Airspeed (knots)**	**Vertical Speed**
	FULL POWER GEAR DOWN FULL FLAPS		65	**MAXIMUM**

118

CLIMB PERFORMANCE AT VARIOUS AIRSPEEDS & DENSITY ALTITUDES IN GO-AROUND CONFIGURATION
(GEAR DOWN/FULL FLAPS/FULL POWER, MAXIMUM GROSS WEIGHT)

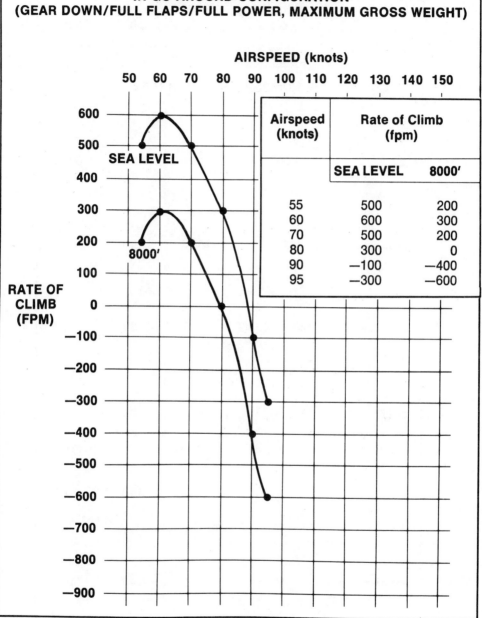

Airspeed (knots)	Rate of Climb (fpm)	
	SEA LEVEL	8000'
55	500	200
60	600	300
70	500	200
80	300	0
90	−100	−400
95	−300	−600

BEECHCRAFT BONANZA   185 HP

ATTITUDE	POWER		PERFORMANCE	
	MAP/Landing Gear	RPM	Airspeed (mph)	Vertical Speed (fpm)
THE NUMBERS FOR BASIC IFR				
	20″	2150	100	500 ↑
	15″	2150	100	0
	15″ + Gear	2150	100	500 ↓
	20″ + Gear	2150	100	0
	20″	2150	140	0
	15″	2150	140	500 ↓

120

ATTITUDE	POWER		PERFORMANCE	
	MAP/Landing Gear	RPM	Airspeed (mph)	Vertical Speed (fpm)

2

THE NUMBERS FOR HIGH-SPEED DESCENT

ATTITUDE	MAP/Landing Gear	RPM	Airspeed (mph)	Vertical Speed (fpm)
	20″	2150	160	500 ↓
	15″	2150	160	1000 ↓

3

LANDING GEAR DRAG AT VARIOUS AIRSPEEDS

Airspeed (mph)	Change in Vertical Speed (fpm)
55	0
60	100
70	200
80	300
90	400
100	500

121

BEECHCRAFT BONANZA   185 HP

4

MAXIMUM SAFE BANK ANGLE DURING GO-AROUND

Angle of Bank	Safety Margin*
0°	22 mph
15°	17 mph
30°	12 mph
45°	2 mph

* This represents the difference between stall speed with gear down/full flaps/full power, and the optimum go-around airspeed of 67 mph.

5

THE NUMBERS FOR MAXIMUM PERFORMANCE (GO-AROUND)

ATTITUDE	POWER		PERFORMANCE	
	MAP/Landing Gear	RPM	Airspeed (mph)	Vertical Speed
4000'				
10000'	FULL POWER GEAR DOWN FULL FLAPS		67	MAXIMUM

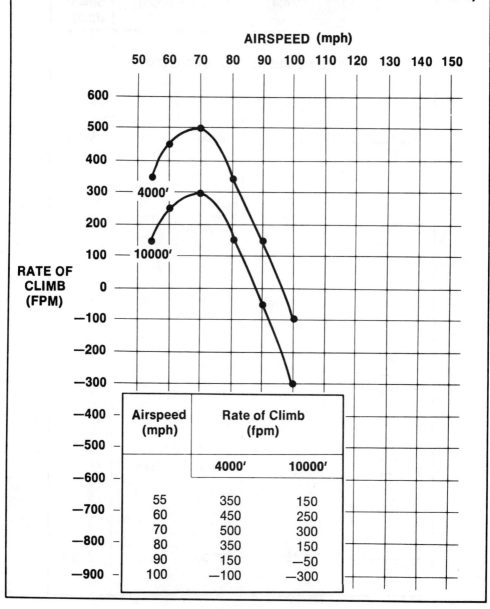

CLIMB PERFORMANCE AT VARIOUS AIRSPEEDS & DENSITY ALTITUDES IN GO-AROUND CONFIGURATION
(GEAR DOWN/FULL FLAPS/FULL POWER, MAXIMUM GROSS WEIGHT)

Airspeed (mph)	Rate of Climb (fpm)	
	4000'	10000'
55	350	150
60	450	250
70	500	300
80	350	150
90	150	−50
100	−100	−300

BEECHCRAFT BONANZA   225 HP

ATTITUDE	POWER		PERFORMANCE	
	MAP/Landing Gear	RPM	Airspeed (mph)	Vertical Speed (fpm)

THE NUMBERS FOR BASIC IFR

ATTITUDE	MAP/Landing Gear	RPM	Airspeed (mph)	Vertical Speed (fpm)
	20″	2300	100	500 ↑
	15″	2300	100	0
	15″ + Gear	2300	100	500 ↓
	20″ + Gear	2300	100	0
	20″	2300	140	0
	15″	2300	140	500 ↓

ATTITUDE	POWER		PERFORMANCE	
	MAP/Landing Gear	RPM	Airspeed (mph)	Vertical Speed (fpm)
THE NUMBERS FOR HIGH-SPEED DESCENT				
▬▬▬	20″	2300	160	500 ↓
▬▬▬	15″	2300	160	1000 ↓

2

3

LANDING GEAR DRAG AT VARIOUS AIRSPEEDS	
Airspeed (mph)	*Change* in Vertical Speed (fpm)
55	0
60	100
70	200
80	300
90	400
100	500

BEECHCRAFT BONANZA   225 HP

4

MAXIMUM SAFE BANK ANGLE DURING GO-AROUND	
Angle of Bank	**Safety Margin***
0°	22 mph
15°	17 mph
30°	12 mph
45°	2 mph

* This represents the difference between stall speed with gear down/full
flaps/full power, and the optimum go-
around airspeed of **67** mph.

5

THE NUMBERS FOR MAXIMUM PERFORMANCE (GO-AROUND)				
ATTITUDE	**POWER**		**PERFORMANCE**	
4000'	**MAP/Landing Gear**	**RPM**	**Airspeed (mph)**	**Vertical Speed**
10000'	**FULL POWER GEAR DOWN FULL FLAPS**		67	**MAXIMUM**

126

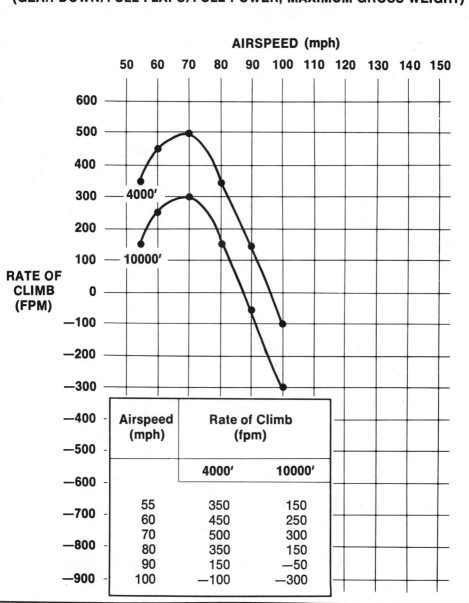

**CLIMB PERFORMANCE AT VARIOUS AIRSPEEDS & DENSITY ALTITUDES IN GO-AROUND CONFIGURATION
(GEAR DOWN/FULL FLAPS/FULL POWER, MAXIMUM GROSS WEIGHT)**

Airspeed (mph)	Rate of Climb (fpm)	
	4000'	**10000'**
55	350	150
60	450	250
70	500	300
80	350	150
90	150	−50
100	−100	−300

BEECHCRAFT BONANZA   285 HP

ATTITUDE	POWER		PERFORMANCE	
	MAP/Landing Gear	RPM	Airspeed (mph)	Vertical Speed (fpm)
	THE NUMBERS FOR BASIC IFR			
	20"	2400	120	500 ↑
	15"	2400	120	0
	15" + Gear	2400	120	500 ↓
	20" + Gear	2400	120	0
	20"	2400	160	0
	15"	2400	160	500 ↓

ATTITUDE	POWER		PERFORMANCE		**2**
	MAP/Landing Gear	RPM	Airspeed (mph)	Vertical Speed (fpm)	

THE NUMBERS FOR HIGH-SPEED DESCENT

ATTITUDE	MAP/Landing Gear	RPM	Airspeed (mph)	Vertical Speed (fpm)
	20″	2400	180	500 ↓
	15″	2400	180	1000 ↓

3

LANDING GEAR DRAG AT VARIOUS AIRSPEEDS

Airspeed (mph)	*Change* in Vertical Speed (fpm)
60	0
70	50
80	100
90	200
100	300
110	400
120	500
140	1000
160	1500

4

MAXIMUM SAFE BANK ANGLE DURING GO-AROUND	
Angle of Bank	**Safety Margin***
0°	27 mph
15°	22 mph
30°	17 mph
45°	7 mph

* This represents the difference between stall speed with gear down/full flaps/full power, and the optimum go-around airspeed of 77 mph.

5

THE NUMBERS FOR MAXIMUM PERFORMANCE (GO-AROUND)				
ATTITUDE	**POWER**		**PERFORMANCE**	
4000'	**MAP/Landing Gear**	**RPM**	**Airspeed (mph)**	**Vertical Speed**
10000'	**FULL POWER GEAR DOWN FULL FLAPS**		77	**MAXIMUM**

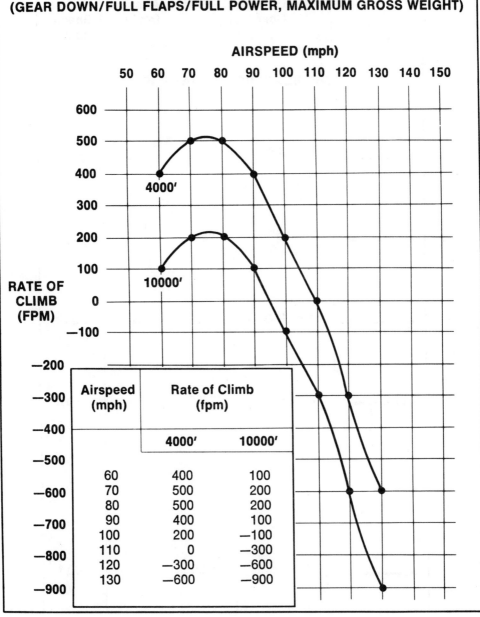

CLIMB PERFORMANCE AT VARIOUS AIRSPEEDS & DENSITY ALTITUDES IN GO-AROUND CONFIGURATION
(GEAR DOWN/FULL FLAPS/FULL POWER, MAXIMUM GROSS WEIGHT)

Airspeed (mph)	Rate of Climb (fpm)	
	4000'	10000'
60	400	100
70	500	200
80	500	200
90	400	100
100	200	−100
110	0	−300
120	−300	−600
130	−600	−900

BEECHCRAFT BARON   260 HP

1

ATTITUDE	POWER		PERFORMANCE	
	MAP/Landing Gear	RPM	Airspeed (mph)	Vertical Speed (fpm)
THE NUMBERS FOR BASIC IFR				
	20″	2450	140	500 ↑
	15″	2450	140	0
	15″ + Gear	2450	140	500 ↓
	20″ + Gear	2450	140	0
	20″	2450	180	0
	15″	2450	180	500 ↓

ATTITUDE	POWER		PERFORMANCE		**2**
	MAP/Landing Gear	RPM	Airspeed (mph)	Vertical Speed (fpm)	
THE NUMBERS FOR HIGH-SPEED DESCENT					
	20″	2450	210	500 ↓	
	15″	2450	210	1000 ↓	

3

LANDING GEAR DRAG AT VARIOUS AIRSPEEDS	
Airspeed (mph)	*Change* in Vertical Speed (fpm)
90	0
100	100
120	200
140	500
160	1000
175	1500

BEECHCRAFT BARON   260 HP

4

MAXIMUM SAFE BANK ANGLE DURING GO-AROUND

Angle of Bank	Safety Margin*
0°	25 mph
15°	20 mph
30°	15 mph
45°	5 mph

* This represents the difference between stall speed with gear down/full flaps/full power, and the optimum go-around airspeed of **80** mph.

5

THE NUMBERS FOR MAXIMUM PERFORMANCE (GO-AROUND)				
ATTITUDE	**POWER**		**PERFORMANCE**	
	MAP/Landing Gear	**RPM**	**Airspeed (mph)**	**Vertical Speed**
SEA LEVEL				
10000'	**FULL POWER GEAR DOWN FULL FLAPS**		80	**MAXIMUM**

6

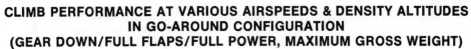

CLIMB PERFORMANCE AT VARIOUS AIRSPEEDS & DENSITY ALTITUDES IN GO-AROUND CONFIGURATION
(GEAR DOWN/FULL FLAPS/FULL POWER, MAXIMUM GROSS WEIGHT)

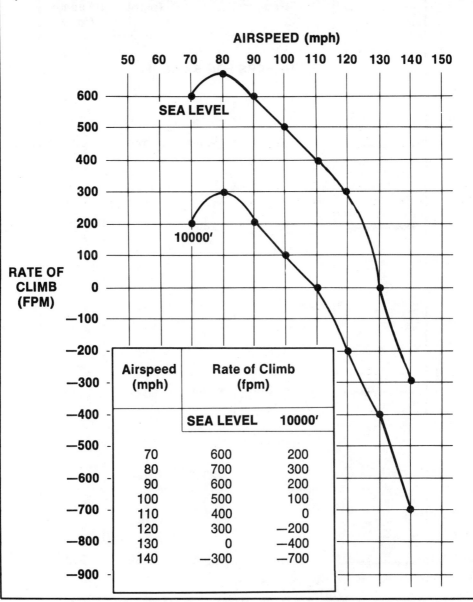

Airspeed (mph)	Rate of Climb (fpm)	
	SEA LEVEL	10000'
70	600	200
80	700	300
90	600	200
100	500	100
110	400	0
120	300	−200
130	0	−400
140	−300	−700

BEECHCRAFT BARON   285 HP

ATTITUDE	POWER		PERFORMANCE	
	MAP/Landing Gear	RPM	Airspeed (mph)	Vertical Speed (fpm)

THE NUMBERS FOR BASIC IFR

ATTITUDE	MAP/Landing Gear	RPM	Airspeed (mph)	Vertical Speed (fpm)
	20″	2300	140	500 ↑
	15″	2300	140	0
	15″ + Gear	2300	140	500 ↓
	20″ + Gear	2300	140	0
	20″	2300	180	0
	15″	2300	180	500 ↓

136

ATTITUDE	POWER		PERFORMANCE	
	MAP/Landing Gear	RPM	Airspeed (mph)	Vertical Speed (fpm)
THE NUMBERS FOR HIGH-SPEED DESCENT				
	20″	2300	210	500 ↓
	15″	2300	210	1000 ↓

2

3

LANDING GEAR DRAG AT VARIOUS AIRSPEEDS

Airspeed (mph)	Change in Vertical Speed (fpm)
90	0
100	100
120	200
140	500
160	1000
175	1500

137

BEECHCRAFT BARON   285 HP

4

MAXIMUM SAFE BANK ANGLE DURING GO-AROUND

Angle of Bank	Safety Margin*
0°	25 mph
15°	20 mph
30°	15 mph
45°	5 mph

* This represents the difference between stall speed with gear down/full flaps/full power, and the optimum go-around airspeed of **80** mph.

5

THE NUMBERS FOR MAXIMUM PERFORMANCE (GO-AROUND)

ATTITUDE	POWER		PERFORMANCE	
	MAP/Landing Gear	RPM	Airspeed (mph)	Vertical Speed
SEA LEVEL				
10000'	FULL POWER GEAR DOWN FULL FLAPS		80	MAXIMUM

138

CLIMB PERFORMANCE AT VARIOUS AIRSPEEDS & DENSITY ALTITUDES IN GO-AROUND CONFIGURATION
(GEAR DOWN/FULL FLAPS/FULL POWER, MAXIMUM GROSS WEIGHT)

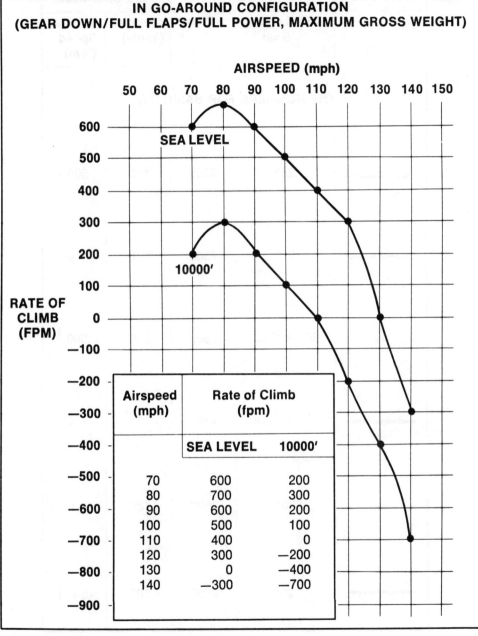

Airspeed (mph)	Rate of Climb (fpm)	
	SEA LEVEL	10000'
70	600	200
80	700	300
90	600	200
100	500	100
110	400	0
120	300	−200
130	0	−400
140	−300	−700

BEECHCRAFT DUKE   380 HP

ATTITUDE	POWER		PERFORMANCE	
	MAP/Landing Gear	RPM	Airspeed (knots)	Vertical Speed (fpm)
THE NUMBERS FOR BASIC IFR				
	25″	2500	140	500 ↑
	20″	2500	140	0
	20″ + Gear	2500	140	500 ↓
	25″ + Gear	2500	140	0
	25″	2500	180	0
	20″	2500	180	500 ↓

140

ATTITUDE	POWER		PERFORMANCE	
	MAP/Landing Gear	RPM	Airspeed (knots)	Vertical Speed (fpm)
THE NUMBERS FOR HIGH-SPEED DESCENT				
	25″	2500	210	500↓
	20″	2500	210	1000↓

LANDING GEAR DRAG AT VARIOUS AIRSPEEDS	
Airspeed (knots)	*Change* in Vertical Speed (fpm)
85	0
90	50
100	100
110	200
140	500

BEECHCRAFT DUKE   380 HP

4

MAXIMUM SAFE BANK ANGLE DURING GO-AROUND	
Angle of Bank	**Safety Margin***
0°	15 knots
15°	10 knots
30°	5 knots
45°	—5 knots

* This represents the difference between stall speed with gear down/full flaps/full power, and the optimum go-around airspeed of **95** knots.

5

THE NUMBERS FOR MAXIMUM PERFORMANCE (GO-AROUND)				
ATTITUDE	**POWER**		**PERFORMANCE**	
	MAP/Landing Gear	**RPM**	**Airspeed (knots)**	**Vertical Speed**
	FULL POWER GEAR DOWN FULL FLAPS		95	**MAXIMUM**

CLIMB PERFORMANCE AT VARIOUS AIRSPEEDS & DENSITY ALTITUDES IN GO-AROUND CONFIGURATION
(GEAR DOWN/FULL FLAPS/FULL POWER, MAXIMUM GROSS WEIGHT)

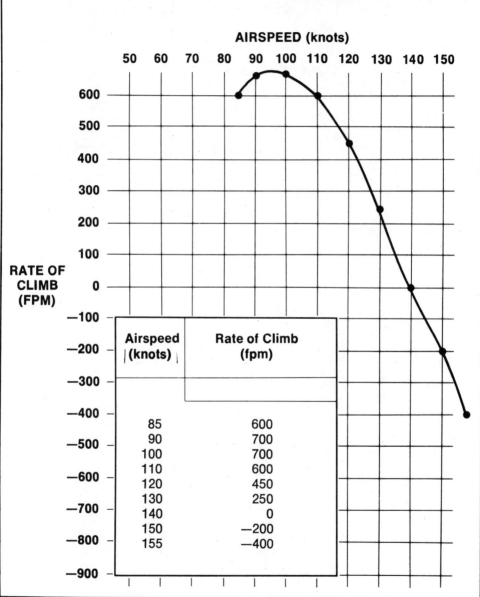

Airspeed (knots)	Rate of Climb (fpm)
85	600
90	700
100	700
110	600
120	450
130	250
140	0
150	−200
155	−400

143

CESSNA 150   100 HP

ATTITUDE	POWER		PERFORMANCE	
	MAP/Landing Gear	RPM	Airspeed (mph)	Vertical Speed (fpm)
THE NUMBERS FOR BASIC IFR				
		2650	80	500 ↑
		2400	90	0
		2100	100	500 ↓
		2650	110	0
		2400	115	500 ↓

ATTITUDE	POWER		PERFORMANCE	
	MAP/Landing Gear	RPM	Airspeed (mph)	Vertical Speed (fpm)

THE NUMBERS FOR HIGH-SPEED DESCENT

ATTITUDE	POWER		PERFORMANCE	
		2650	120	500 ↓
		2400	125	1000 ↓

CESSNA 150   100 HP

4

MAXIMUM SAFE BANK ANGLE DURING GO-AROUND

Angle of Bank	Safety Margin*
0°	15 mph
15°	10 mph
30°	5 mph
45°	−5 mph

* This represents the difference between stall speed with gear down/full flaps/full power, and the optimum go-around airspeed of 50 mph.

5

THE NUMBERS FOR MAXIMUM PERFORMANCE (GO-AROUND)

ATTITUDE	POWER		PERFORMANCE	
	MAP/Landing Gear	RPM	Airspeed (mph)	Vertical Speed
	FULL POWER FULL FLAPS		50	MAXIMUM

146

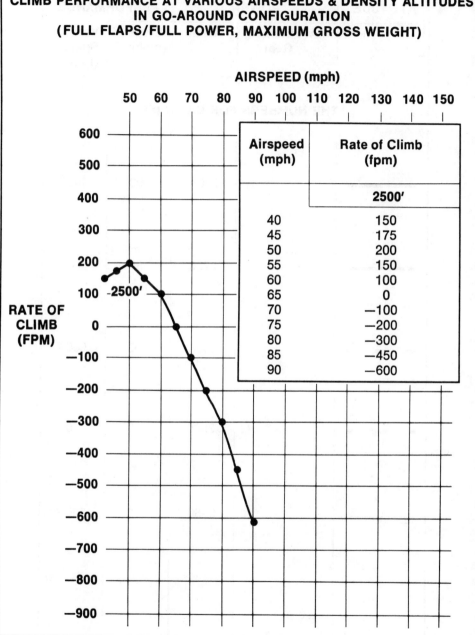

CLIMB PERFORMANCE AT VARIOUS AIRSPEEDS & DENSITY ALTITUDES
IN GO-AROUND CONFIGURATION
(FULL FLAPS/FULL POWER, MAXIMUM GROSS WEIGHT)

Airspeed (mph)	Rate of Climb (fpm)
	2500'
40	150
45	175
50	200
55	150
60	100
65	0
70	−100
75	−200
80	−300
85	−450
90	−600

147

CESSNA 172   150 HP

1

ATTITUDE	POWER		PERFORMANCE	
	MAP/Landing Gear	RPM	Airspeed (mph)	Vertical Speed (fpm)

THE NUMBERS FOR BASIC IFR

ATTITUDE	MAP/Landing Gear	RPM	Airspeed (mph)	Vertical Speed (fpm)
		2500	90	500 ↑
		2250	95	0
		2000	100	500 ↓
		2500	110	0
		2250	115	500 ↓

ATTITUDE	POWER		PERFORMANCE	
	MAP/Landing Gear	RPM	Airspeed (mph)	Vertical Speed (fpm)
THE NUMBERS FOR HIGH-SPEED DESCENT				
		2500	120	500 ↓
		2250	125	1000 ↓

4

MAXIMUM SAFE BANK ANGLE DURING GO-AROUND	
Angle of Bank	**Safety Margin***
0°	17 mph
15°	12 mph
30°	7 mph
45°	—3 mph

* This represents the difference between stall speed with gear down/full flaps/full power, and the optimum go-around airspeed of **55** mph.

5

THE NUMBERS FOR MAXIMUM PERFORMANCE (GO-AROUND)				
ATTITUDE	**POWER**		**PERFORMANCE**	
SEA LEVEL	**MAP/Landing Gear**	**RPM**	**Airspeed (mph)**	**Vertical Speed**
6000′	**FULL POWER FULL FLAPS**		55	**MAXIMUM**

6

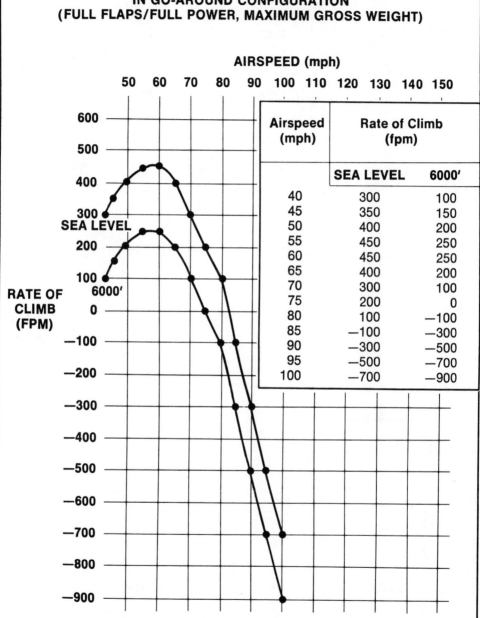

CLIMB PERFORMANCE AT VARIOUS AIRSPEEDS & DENSITY ALTITUDES
IN GO-AROUND CONFIGURATION
(FULL FLAPS/FULL POWER, MAXIMUM GROSS WEIGHT)

Airspeed (mph)	Rate of Climb (fpm)	
	SEA LEVEL	6000'
40	300	100
45	350	150
50	400	200
55	450	250
60	450	250
65	400	200
70	300	100
75	200	0
80	100	−100
85	−100	−300
90	−300	−500
95	−500	−700
100	−700	−900

151

CESSNA CARDINAL RG   200 HP

ATTITUDE	POWER		PERFORMANCE	
	MAP/Landing Gear	RPM	Airspeed (knots)	Vertical Speed (fpm)
	THE NUMBERS FOR BASIC IFR			
	22"	2400	90	500 ↑
	17"	2400	95	0
	17" + Gear	2400	100	500 ↓
	22" + Gear	2400	95	0
	22"	2400	120	0
	18"	2400	125	500 ↓

2

ATTITUDE	POWER		PERFORMANCE	
	MAP/Landing Gear	RPM	Airspeed (knots)	Vertical Speed (fpm)

THE NUMBERS FOR HIGH-SPEED DESCENT

ATTITUDE	MAP/Landing Gear	RPM	Airspeed (knots)	Vertical Speed (fpm)
	22"	2400	135	500 ↓
	17"	2400	137	1000 ↓

3

LANDING GEAR DRAG AT VARIOUS AIRSPEEDS

Airspeed (knots)	Change in Vertical Speed (fpm)
60	0
70	100
80	300
90	400
100	500
110	700
120	900
125	1000

153

4

MAXIMUM SAFE BANK ANGLE DURING GO-AROUND	
Angle of Bank	**Safety Margin***
0°	17 knots
15°	12 knots
30°	7 knots
45°	—10 knots

* This represents the difference between stall speed with gear down/full flaps/full power, and the optimum go-around airspeed of 60 knots.

5

THE NUMBERS FOR MAXIMUM PERFORMANCE (GO-AROUND)				
ATTITUDE	**POWER**		**PERFORMANCE**	
	MAP/Landing Gear	RPM	Airspeed (knots)	Vertical Speed
	FULL POWER GEAR DOWN FULL FLAPS		60	MAXIMUM

CLIMB PERFORMANCE AT VARIOUS AIRSPEEDS & DENSITY ALTITUDES IN GO-AROUND CONFIGURATION
(GEAR DOWN/FULL FLAPS/FULL POWER, MAXIMUM GROSS WEIGHT)

AIRSPEED (knots)

Airspeed (knots)	Rate of Climb (fpm)
	6000'
50	400
60	500
70	300
80	50
90	—200
95	—350

RATE OF CLIMB (FPM)

155

CESSNA SKYLANE   235 HP

1

ATTITUDE	POWER		PERFORMANCE	
	MAP/Landing Gear	RPM	Airspeed (mph)	Vertical Speed (fpm)
THE NUMBERS FOR BASIC IFR				
	23″	2350	95	500 ↑
	18″	2350	100	0
	14″	2350.	105	500 ↓
	23″	2350	130	0
	18″	2350	135	500 ↓

2

ATTITUDE	POWER		PERFORMANCE	
	MAP/Landing Gear	RPM	Airspeed (mph)	Vertical Speed (fpm)
THE NUMBERS FOR HIGH-SPEED DESCENT				
	23"	2350	140	500 ↓
	18"	2350	145	1000 ↓

CESSNA SKYLANE   235 HP

4

MAXIMUM SAFE BANK ANGLE DURING GO-AROUND	
Angle of Bank	**Safety Margin***
0°	16 mph
15°	11 mph
30°	6 mph
45°	—5 mph

* This represents the difference between stall speed with gear down/full flaps/full power, and the optimum go-around airspeed of **70** mph.

5

THE NUMBERS FOR MAXIMUM PERFORMANCE (GO-AROUND)				
ATTITUDE	**POWER**		**PERFORMANCE**	
	MAP/Landing Gear	**RPM**	**Airspeed (mph)**	**Vertical Speed**
	FULL POWER FULL FLAPS		70	**MAXIMUM**

158

CLIMB PERFORMANCE AT VARIOUS AIRSPEEDS & DENSITY ALTITUDES IN GO-AROUND CONFIGURATION
(GEAR DOWN/FULL FLAPS/FULL POWER, MAXIMUM GROSS WEIGHT)

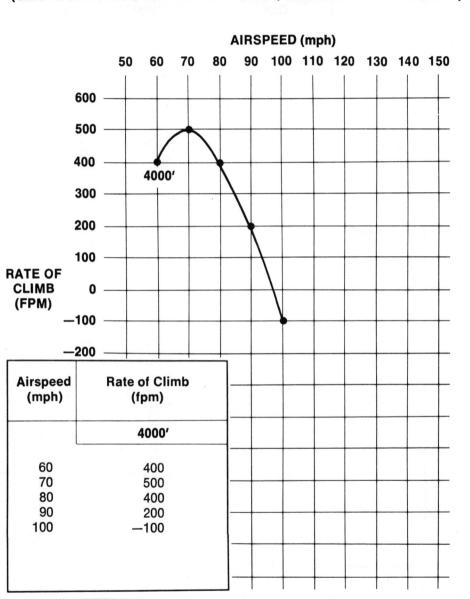

Airspeed (mph)	Rate of Climb (fpm)
	4000'
60	400
70	500
80	400
90	200
100	−100

159

CESSNA CENTURION II   285 HP

ATTITUDE	POWER		PERFORMANCE	
	MAP/Landing Gear	RPM	Airspeed (mph)	Vertical Speed (fpm)

THE NUMBERS FOR BASIC IFR

ATTITUDE	MAP/Landing Gear	RPM	Airspeed (mph)	Vertical Speed (fpm)
	20"	2400	140	500 ↑
	15"	2400	140	0
	15" + Gear	2400	145	500 ↓
	20" + Gear	2400	140	0
	20"	2400	160	0
	15"	2400	165	500 ↓

ATTITUDE	POWER		PERFORMANCE		**2**
	MAP/Landing Gear	RPM	Airspeed (mph)	Vertical Speed (fpm)	
	THE NUMBERS FOR HIGH-SPEED DESCENT				
	20″	2400	175	500 ↓	
	15″	2400	175	1000 ↓	

3

LANDING GEAR DRAG AT VARIOUS AIRSPEEDS	
Airspeed (mph)	*Change* in Vertical Speed (fpm)
80	0
90	50
100	100
110	200
120	300
130	400
140	500
150	700

161

CESSNA CENTURION II   285 HP

4

MAXIMUM SAFE BANK ANGLE DURING GO-AROUND	
Angle of Bank	**Safety Margin***
0°	20 mph
15°	15 mph
30°	10 mph
45°	5 mph

* This represents the difference between stall speed with gear down/full flaps/full power, and the optimum go-around airspeed of **65** mph.

5

THE NUMBERS FOR MAXIMUM PERFORMANCE (GO-AROUND)				
ATTITUDE	**POWER**		**PERFORMANCE**	
	MAP/Landing Gear	**RPM**	**Airspeed (mph)**	**Vertical Speed**
	FULL POWER GEAR DOWN FULL FLAPS		65	MAXIMUM

CLIMB PERFORMANCE AT VARIOUS AIRSPEEDS & DENSITY ALTITUDES IN GO-AROUND CONFIGURATION
(GEAR DOWN/FULL FLAPS/FULL POWER, MAXIMUM GROSS WEIGHT)

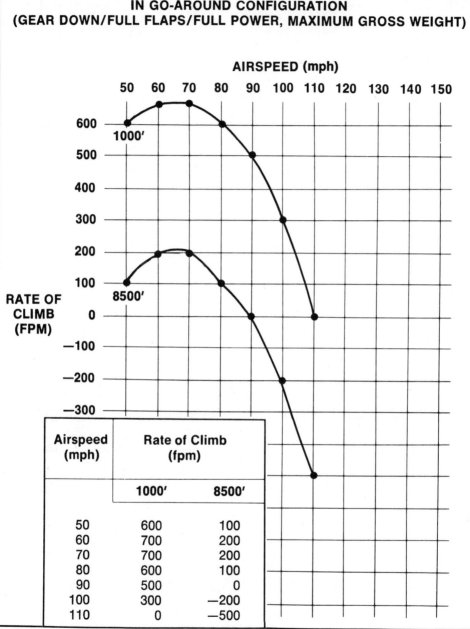

Airspeed (mph)	Rate of Climb (fpm)	
	1000′	8500′
50	600	100
60	700	200
70	700	200
80	600	100
90	500	0
100	300	−200
110	0	−500

163

CESSNA SKYMASTER   210 HP

ATTITUDE	POWER		PERFORMANCE	
	MAP/Landing Gear	RPM	Airspeed (mph)	Vertical Speed (fpm)

THE NUMBERS FOR BASIC IFR

ATTITUDE	MAP/Landing Gear	RPM	Airspeed (mph)	Vertical Speed (fpm)
	20″	2450	140	500 ↑
	15″	2450	130	0
	15″ + Gear	2450	140	500 ↓
	20″ + Gear	2450	130	0
	20″	2450	160	0
	15″	2450	160	500 ↓

ATTITUDE	POWER		PERFORMANCE	
	MAP/Landing Gear	RPM	Airspeed (mph)	Vertical Speed (fpm)
THE NUMBERS FOR HIGH-SPEED DESCENT				
	20″	2450	180	500 ↓
	15″	2450	180	1000 ↓

LANDING GEAR DRAG AT VARIOUS AIRSPEEDS

Airspeed (mph)	*Change* in Vertical Speed (fpm)
90	0
100	50
110	100
120	200
130	400
140	500

CESSNA SKYMASTER   210 HP

4

MAXIMUM SAFE BANK ANGLE DURING GO-AROUND	
Angle of Bank	**Safety Margin***
0°	25 mph
15°	20 mph
30°	15 mph
45°	5 mph

* This represents the difference between stall speed with gear down/full flaps/full power, and the optimum go-around airspeed of 95 mph.

5

THE NUMBERS FOR MAXIMUM PERFORMANCE (GO-AROUND)				
ATTITUDE	**POWER**		**PERFORMANCE**	
	MAP/Landing Gear	**RPM**	**Airspeed (mph)**	**Vertical Speed**
	FULL POWER GEAR DOWN FULL FLAPS		95	MAXIMUM

CLIMB PERFORMANCE AT VARIOUS AIRSPEEDS & DENSITY ALTITUDES IN GO-AROUND CONFIGURATION
(GEAR DOWN/FULL FLAPS/FULL POWER, MAXIMUM GROSS WEIGHT)

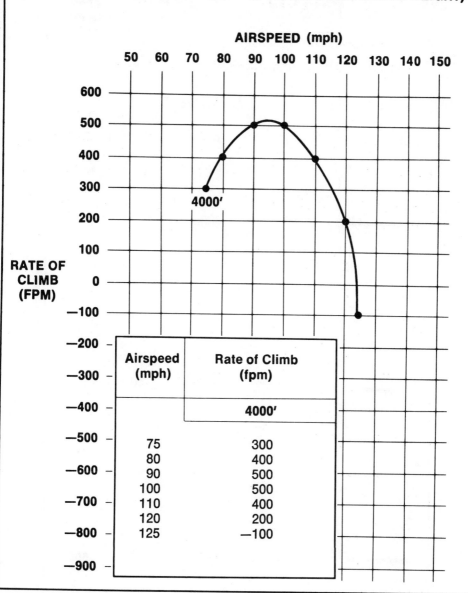

Airspeed (mph)	Rate of Climb (fpm)
	4000'
75	300
80	400
90	500
100	500
110	400
120	200
125	−100

167

CESSNA 310   285 HP

ATTITUDE	POWER		PERFORMANCE	
	MAP/Landing Gear	RPM	Airspeed (mph)	Vertical Speed (fpm)

THE NUMBERS FOR BASIC IFR

ATTITUDE	MAP/Landing Gear	RPM	Airspeed (mph)	Vertical Speed (fpm)
	20″	2300	140	500 ↑
	15″	2300	140	0
	15″ + Gear	2300	140	500 ↓
	20″ + Gear	2300	140	0
	20″	2300	180	0
	15″	2300	180	500 ↓

ATTITUDE	POWER		PERFORMANCE	
	MAP/Landing Gear	RPM	Airspeed (mph)	Vertical Speed (fpm)
THE NUMBERS FOR HIGH-SPEED DESCENT				
	20″	2300	200	500 ↓
	15″	2300	200	1000 ↓

LANDING GEAR DRAG AT VARIOUS AIRSPEEDS	
Airspeed (mph)	*Change* in Vertical Speed (fpm)
90	0
100	100
110	200
120	300
130	400
140	500

CESSNA 310   285 HP

4

MAXIMUM SAFE BANK ANGLE DURING GO-AROUND	
Angle of Bank	**Safety Margin***
0°	20 mph
15°	15 mph
30°	10 mph
45°	0 mph

* This represents the difference between stall speed with gear down/full
flaps/full power, and the optimum go-
around airspeed of **95** mph.

5

THE NUMBERS FOR MAXIMUM PERFORMANCE (GO-AROUND)				
ATTITUDE	**POWER**		**PERFORMANCE**	
	MAP/Landing Gear	RPM	Airspeed (mph)	Vertical Speed
⊂━○━⊃	FULL POWER GEAR DOWN FULL FLAPS		95	MAXIMUM

170

CLIMB PERFORMANCE AT VARIOUS AIRSPEEDS & DENSITY ALTITUDES IN GO-AROUND CONFIGURATION
(GEAR DOWN/FULL FLAPS/FULL POWER, MAXIMUM GROSS WEIGHT)

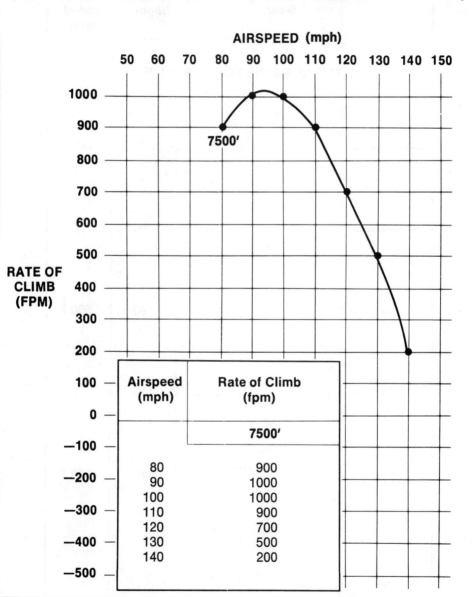

Airspeed (mph)	Rate of Climb (fpm)
	7500′
80	900
90	1000
100	1000
110	900
120	700
130	500
140	200

GRUMMAN AMERICAN AA5   150 HP

ATTITUDE	POWER		PERFORMANCE	
	MAP/Landing Gear	RPM	Airspeed (mph)	Vertical Speed (fpm)

THE NUMBERS FOR BASIC IFR

ATTITUDE	MAP/Landing Gear	RPM	Airspeed (mph)	Vertical Speed (fpm)
		2250	90	500 ↑
		2000	90	0
		1700	90	500 ↓
		2250	110	0
		1700	110	500 ↓

2

ATTITUDE	POWER		PERFORMANCE	
	MAP/Landing Gear	RPM	Airspeed (mph)	Vertical Speed (fpm)
THE NUMBERS FOR HIGH-SPEED DESCENT				
		2250	125	500 ↓
		2000	125	1000 ↓

4

MAXIMUM SAFE BANK ANGLE DURING GO-AROUND	
Angle of Bank	**Safety Margin***
0°	20 mph
15°	15 mph
30°	10 mph
45°	0 mph

* This represents the difference between stall speed with gear down/full flaps/full power, and the optimum go-around airspeed of **70** mph.

5

THE NUMBERS FOR MAXIMUM PERFORMANCE (GO-AROUND)				
ATTITUDE	**POWER**		**PERFORMANCE**	
	MAP/Landing Gear	**RPM**	**Airspeed (mph)**	**Vertical Speed**
	FULL POWER FULL FLAPS		70	MAXIMUM

CLIMB PERFORMANCE AT VARIOUS AIRSPEEDS & DENSITY ALTITUDES IN GO-AROUND CONFIGURATION
(FULL FLAPS/FULL POWER, MAXIMUM GROSS WEIGHT)

AIRSPEED (mph)

50 60 70 80 90 100 110 120 130 140 150

RATE OF CLIMB (FPM)

600
500
400
300
200
100
0
−100
−200
−300
−400
−500
−600
−700
−800
−900

Airspeed (mph)	Rate of Climb (fpm)
	5000'
60	500
70	600
80	400
90	250
100	−100
110	−500

175

GRUMMAN AMERICAN TIGER   180 HP

ATTITUDE	POWER		PERFORMANCE	
	MAP/Landing Gear	RPM	Airspeed (knots)	Vertical Speed (fpm)
THE NUMBERS FOR BASIC IFR				
		2400	100	500 ↑
		2200	100	0
		2000	100	500 ↓
		2400	120	0
		2200	120	500 ↓

2

ATTITUDE	POWER		PERFORMANCE	
	MAP/Landing Gear	RPM	Airspeed (knots)	Vertical Speed (fpm)

THE NUMBERS FOR HIGH-SPEED DESCENT

ATTITUDE	MAP/Landing Gear	RPM	Airspeed (knots)	Vertical Speed (fpm)
▬		2400	130	500 ↓
▭		2200	130	1000 ↓

177

4

MAXIMUM SAFE BANK ANGLE DURING GO-AROUND	
Angle of Bank	**Safety Margin***
0°	15 knots
15°	10 knots
30°	5 knots
45°	—5 knots

* This represents the difference between stall speed with gear down/full flaps/full power, and the optimum go-around airspeed of 60 knots.

5

THE NUMBERS FOR MAXIMUM PERFORMANCE (GO-AROUND)				
ATTITUDE	**POWER**		**PERFORMANCE**	
Sea Level	**MAP/Landing Gear**	**RPM**	**Airspeed (knots)**	**Vertical Speed**
8000'	FULL POWER FULL FLAPS		60	MAXIMUM

CLIMB PERFORMANCE AT VARIOUS AIRSPEEDS & DENSITY ALTITUDES IN GO-AROUND CONFIGURATION
(FULL FLAPS/FULL POWER, MAXIMUM GROSS WEIGHT)

Airspeed (knots)	Rate of Climb (fpm)	
	Sea Level	8000'
50	500	100
60	650	250
70	500	100
80	300	−100
90	50	−350
100	−200	−600

179

PIPER CHEROKEE   150 HP

ATTITUDE	POWER		PERFORMANCE	
	MAP/Landing Gear	RPM	Airspeed (mph)	Vertical Speed (fpm)

THE NUMBERS FOR BASIC IFR

ATTITUDE	POWER		PERFORMANCE	
		2400	90	500 ↑
		2100	90	0
		1800	90	500 ↓
		2400	110	0
		2100	110	500 ↓

2

ATTITUDE	POWER		PERFORMANCE	
	MAP/Landing Gear	RPM	Airspeed (mph)	Vertical Speed (fpm)
THE NUMBERS FOR HIGH-SPEED DESCENT				
		2400	120	500 ↓
		2100	120	1000 ↓

181

PIPER CHEROKEE   150 HP

4

MAXIMUM SAFE BANK ANGLE DURING GO-AROUND

Angle of Bank	Safety Margin*
0°	22 mph
15°	17 mph
30°	12 mph
45°	2 mph

* This represents the difference between stall speed with gear down/full flaps/full power, and the optimum go-around airspeed of 65 mph.

5

THE NUMBERS FOR MAXIMUM PERFORMANCE (GO-AROUND)

ATTITUDE	POWER		PERFORMANCE	
	MAP/Landing Gear	RPM	Airspeed (mph)	Vertical Speed
⌐ ∘ ⌐	FULL POWER GEAR DOWN FULL FLAPS		65	MAXIMUM

182

CLIMB PERFORMANCE AT VARIOUS AIRSPEEDS & DENSITY ALTITUDES IN GO-AROUND CONFIGURATION
(FULL FLAPS/FULL POWER, MAXIMUM GROSS WEIGHT)

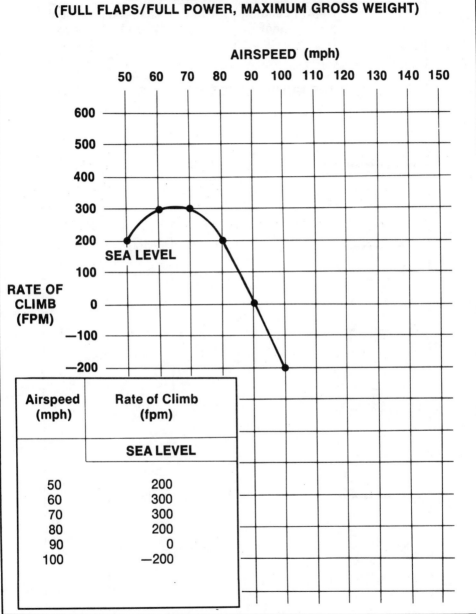

Airspeed (mph)	Rate of Climb (fpm)
	SEA LEVEL
50	200
60	300
70	300
80	200
90	0
100	−200

183

PIPER ARCHER II   180 HP

ATTITUDE	POWER		PERFORMANCE	
	MAP/Landing Gear	RPM	Airspeed (knots)	Vertical Speed (fpm)
THE NUMBERS FOR BASIC IFR				
		2300	100	500 ↑
		2000	100	0
		1700	100	500 ↓
		2300	120	0
		1700	120	500 ↓

ATTITUDE	POWER		PERFORMANCE	
	MAP/Landing Gear	RPM	Airspeed (knots)	Vertical Speed (fpm)
THE NUMBERS FOR HIGH-SPEED DESCENT				
		2300	130	500 ↓
		2000	130	1000 ↓

PIPER ARCHER II   180 HP

4

MAXIMUM SAFE BANK ANGLE DURING GO-AROUND	
Angle of Bank	**Safety Margin***
0°	20 knots
15°	15 knots
30°	10 knots
45°	0 knots

* This represents the difference between stall speed with gear down/full flaps/full power, and the optimum go-around airspeed of 60 knots.

5

THE NUMBERS FOR MAXIMUM PERFORMANCE (GO-AROUND)				
ATTITUDE	**POWER**		**PERFORMANCE**	
	MAP/Landing Gear	**RPM**	**Airspeed (knots)**	**Vertical Speed**
⌐⚬⌐	FULL POWER FULL FLAPS		60	MAXIMUM

186

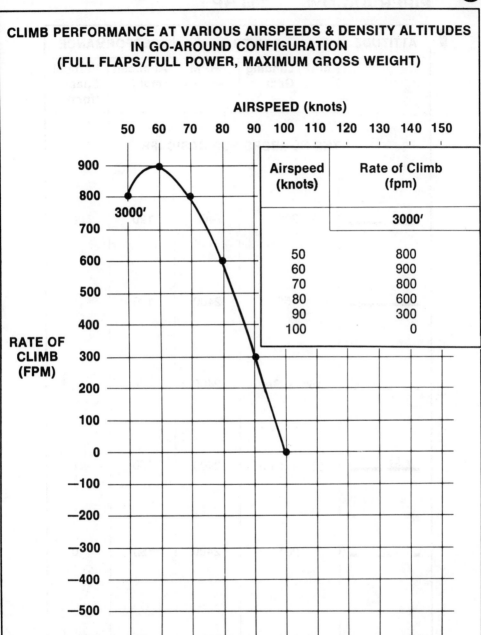

CLIMB PERFORMANCE AT VARIOUS AIRSPEEDS & DENSITY ALTITUDES
IN GO-AROUND CONFIGURATION
(FULL FLAPS/FULL POWER, MAXIMUM GROSS WEIGHT)

AIRSPEED (knots)

Airspeed (knots)	Rate of Climb (fpm)
	3000'
50	800
60	900
70	800
80	600
90	300
100	0

RATE OF CLIMB (FPM)

PIPER ARROW   200 HP

ATTITUDE	POWER		PERFORMANCE	
	MAP/Landing Gear	RPM	Airspeed (mph)	Vertical Speed (fpm)
THE NUMBERS FOR BASIC IFR				
	20"	2400	120	500 ↑
	15"	2400	120	0
	15" + Gear	2400	120	500 ↓
	20" + Gear	2400	120	0
	20"	2400	150	0
	15"	2400	150	500 ↓

ATTITUDE	POWER		PERFORMANCE		**2**
	MAP/Landing Gear	RPM	Airspeed (mph)	Vertical Speed (fpm)	

THE NUMBERS FOR HIGH-SPEED DESCENT

ATTITUDE	MAP/Landing Gear	RPM	Airspeed (mph)	Vertical Speed (fpm)
	20"	2400	165	500 ↓
	15"	2400	165	1000 ↓

3

LANDING GEAR DRAG AT VARIOUS AIRSPEEDS

Airspeed (mph)	*Change* in Vertical Speed (fpm)
60	0
70	50
80	100
90	200
100	300
110	400
120	500

PIPER ARROW   200 HP

4

MAXIMUM SAFE BANK ANGLE DURING GO-AROUND

Angle of Bank	Safety Margin*
0°	20 mph
15°	15 mph
30°	10 mph
45°	0 mph

* This represents the difference between stall speed with gear down/full flaps/full power, and the optimum go-around airspeed of 75 mph.

5

THE NUMBERS FOR MAXIMUM PERFORMANCE (GO-AROUND)

ATTITUDE	POWER		PERFORMANCE	
SEA LEVEL	MAP/Landing Gear	RPM	Airspeed (mph)	Vertical Speed
5000'	FULL POWER GEAR DOWN FULL FLAPS		75	MAXIMUM

190

CLIMB PERFORMANCE AT VARIOUS AIRSPEEDS & DENSITY ALTITUDES IN GO-AROUND CONFIGURATION
(GEAR DOWN/FULL FLAPS/FULL POWER, MAXIMUM GROSS WEIGHT)

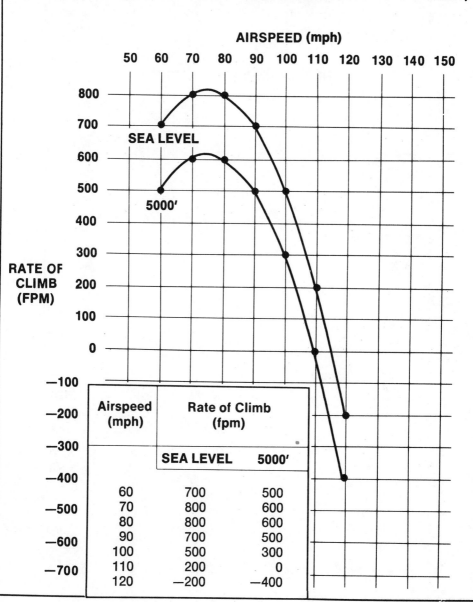

Airspeed (mph)	Rate of Climb (fpm)	
	SEA LEVEL	5000'
60	700	500
70	800	600
80	800	600
90	700	500
100	500	300
110	200	0
120	−200	−400

PIPER COMANCHE 260 HP

ATTITUDE	POWER		PERFORMANCE	
	MAP/Landing Gear	RPM	Airspeed (mph)	Vertical Speed (fpm)
THE NUMBERS FOR BASIC IFR				
	23″	2400	120	500 ↑
	18″	2400	120	0
	18″ + Gear	2400	120	500 ↓
	23″ + Gear	2400	120	0
	23″	2400	150	0
	18″	2400	150	500 ↓

ATTITUDE	POWER		PERFORMANCE		**2**
	MAP/Landing Gear	**RPM**	**Airspeed (mph)**	**Vertical Speed (fpm)**

THE NUMBERS FOR HIGH-SPEED DESCENT

ATTITUDE	MAP/Landing Gear	RPM	Airspeed (mph)	Vertical Speed (fpm)
▬	23″	2400	165	500 ↓
▬	18″	2400	165	1000 ↓

3

**LANDING GEAR DRAG
AT VARIOUS AIRSPEEDS**

Airspeed (mph)	*Change* in Vertical Speed (fpm)
60	0
70	50
80	100
90	200
100	300
110	400
120	500
140	1000

PIPER COMANCHE   260 HP

4

MAXIMUM SAFE BANK ANGLE DURING GO-AROUND	
Angle of Bank	**Safety Margin***
0°	20 mph
15°	15 mph
30°	10 mph
45°	0 mph

* This represents the difference between stall speed with gear down/full
flaps/full power, and the optimum go-
around airspeed of **75** mph.

5

THE NUMBERS FOR MAXIMUM PERFORMANCE (GO-AROUND)				
ATTITUDE	**POWER**		**PERFORMANCE**	
	MAP/Landing Gear	**RPM**	**Airspeed (mph)**	**Vertical Speed**
	FULL POWER GEAR DOWN FULL FLAPS		75	**MAXIMUM**

194

CLIMB PERFORMANCE AT VARIOUS AIRSPEEDS & DENSITY ALTITUDES IN GO-AROUND CONFIGURATION
(GEAR DOWN/FULL FLAPS/FULL POWER, MAXIMUM GROSS WEIGHT)

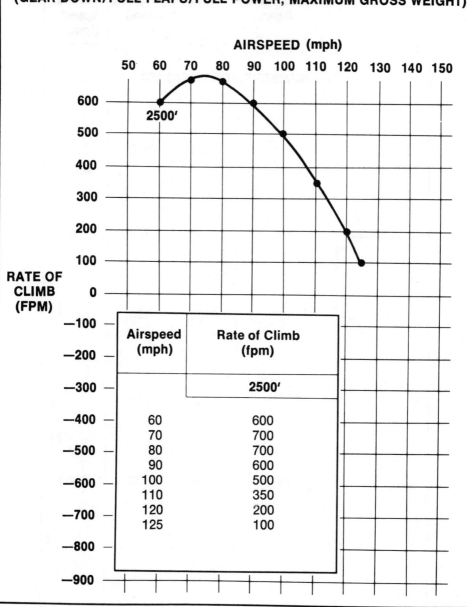

Airspeed (mph)	Rate of Climb (fpm)
	2500'
60	600
70	700
80	700
90	600
100	500
110	350
120	200
125	100

PIPER CHEROKEE SIX   300 HP

1

ATTITUDE	POWER		PERFORMANCE	
	MAP/Landing Gear	RPM	Airspeed (knots)	Vertical Speed (fpm)

THE NUMBERS FOR BASIC IFR

ATTITUDE	MAP/Landing Gear	RPM	Airspeed (knots)	Vertical Speed (fpm)
	23″	2400	100	500 ↑
	18″	2400	100	0
	15″	2400	100	500 ↓
	23″	2400	130	0
	18″	2400	130	500 ↓

ATTITUDE	POWER		PERFORMANCE	
	MAP/Landing Gear	RPM	Airspeed (knots)	Vertical Speed (fpm)

THE NUMBERS FOR HIGH-SPEED DESCENT

ATTITUDE	MAP/Landing Gear	RPM	Airspeed (knots)	Vertical Speed (fpm)
	23"	2400	140	500 ↓
	18"	2400	140	1000 ↓

PIPER CHEROKEE SIX   300 HP

4

MAXIMUM SAFE BANK ANGLE DURING GO-AROUND	
Angle of Bank	**Safety Margin***
0°	15 knots
15°	10 knots
30°	5 knots
45°	0 knots

* This represents the difference between stall speed with gear down/full flaps/full power, and the optimum go-around airspeed of **60** knots.

5

THE NUMBERS FOR MAXIMUM PERFORMANCE (GO-AROUND)				
ATTITUDE	**POWER**		**PERFORMANCE**	
	MAP/Landing Gear	**RPM**	**Airspeed (knots)**	**Vertical Speed**
⎯⎯⊂⚬⊃⎯⎯	**FULL POWER FULL FLAPS**		60	**MAXIMUM**

CLIMB PERFORMANCE AT VARIOUS AIRSPEEDS & DENSITY ALTITUDES IN GO-AROUND CONFIGURATION
(FULL FLAPS/FULL POWER, MAXIMUM GROSS WEIGHT)

AIRSPEED (knots)

Airspeed (knots)	Rate of Climb (fpm)	
	3000'	9000'
50	500	100
60	700	300
70	500	100
80	300	−100
90	100	−300
100	−100	−500
110	−400	−800

RATE OF CLIMB (FPM)

PIPER LANCE 300 HP

ATTITUDE	POWER		PERFORMANCE	
	MAP/Landing Gear	RPM	Airspeed (knots)	Vertical Speed (fpm)
	THE NUMBERS FOR BASIC IFR			
	22"	2400	100	500 ↑
	17"	2400	100	0
	15" + Gear	2400	100	500 ↓
	22" + Gear	2400	100	0
	22"	2400	130	0
	15"	2400	130	500 ↓

ATTITUDE	POWER		PERFORMANCE		**2**
	MAP/Landing Gear	RPM	Airspeed (knots)	Vertical Speed (fpm)

THE NUMBERS FOR HIGH-SPEED DESCENT

ATTITUDE	MAP/Landing Gear	RPM	Airspeed (knots)	Vertical Speed (fpm)
	20″	2400	145	500 ↓
	15″	2400	145	1000 ↓

3

LANDING GEAR DRAG AT VARIOUS AIRSPEEDS

Airspeed (knots)	*Change* in Vertical Speed (fpm)
60	0
70	50
80	100
90	300
100	500
110	700
120	1000

201

PIPER LANCE   300 HP

4

MAXIMUM SAFE BANK ANGLE DURING GO-AROUND

Angle of Bank	Safety Margin*
0°	15 knots
15°	10 knots
30°	5 knots
45°	0 knots

* This represents the difference between stall speed with gear down/full flaps/full power, and the optimum go-around airspeed of 60 knots.

5

THE NUMBERS FOR MAXIMUM PERFORMANCE (GO-AROUND)

ATTITUDE	POWER		PERFORMANCE	
	MAP/Landing Gear	RPM	Airspeed (knots)	Vertical Speed
	FULL POWER GEAR DOWN FULL FLAPS		60	MAXIMUM

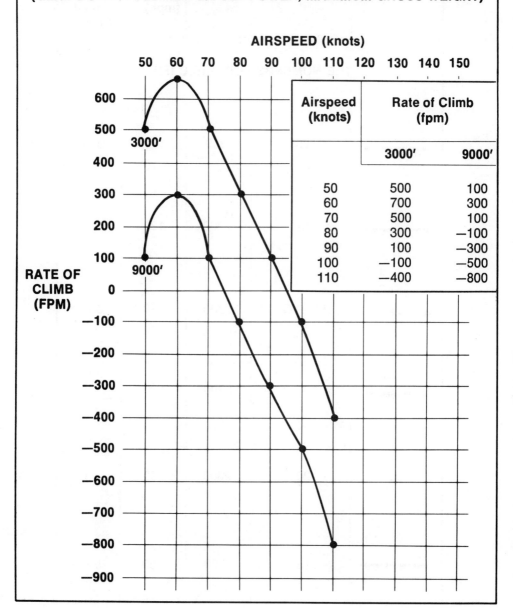

CLIMB PERFORMANCE AT VARIOUS AIRSPEEDS & DENSITY ALTITUDES
IN GO-AROUND CONFIGURATION
(GEAR DOWN/FULL FLAPS/FULL POWER, MAXIMUM GROSS WEIGHT)

AIRSPEED (knots)

Airspeed (knots)	Rate of Climb (fpm)	
	3000'	9000'
50	500	100
60	700	300
70	500	100
80	300	−100
90	100	−300
100	−100	−500
110	−400	−800

PIPER AZTEC   250 HP

ATTITUDE	POWER		PERFORMANCE	
	MAP/Landing Gear	RPM	Airspeed (mph)	Vertical Speed (fpm)

THE NUMBERS FOR BASIC IFR

ATTITUDE	MAP/Landing Gear	RPM	Airspeed (mph)	Vertical Speed (fpm)
	20″	2300	120	500 ↑
	15″	2300	120	0
	15″ + Gear	2300	120	500 ↓
	20″ + Gear	2300	120	0
	20″	2300	150	0
	15″	2300	150	500 ↓

204

ATTITUDE	POWER		PERFORMANCE		**2**
	MAP/Landing Gear	RPM	Airspeed (mph)	Vertical Speed (fpm)	

THE NUMBERS FOR HIGH-SPEED DESCENT

ATTITUDE	MAP/Landing Gear	RPM	Airspeed (mph)	Vertical Speed (fpm)
	20″	2300	165	500 ↓
	15″	2300	165	1000 ↓

3

LANDING GEAR DRAG AT VARIOUS AIRSPEEDS

Airspeed (mph)	*Change* in Vertical Speed (fpm)
70	0
80	100
90	200
100	300
110	400
120	500

PIPER AZTEC   250 HP

4

MAXIMUM SAFE BANK ANGLE DURING GO-AROUND

Angle of Bank	Safety Margin*
0°	18 mph
15°	13 mph
30°	8 mph
45°	—3 mph

* This represents the difference between stall speed with gear down/full flaps/full power, and the optimum go-around airspeed of 80 mph.

5

THE NUMBERS FOR MAXIMUM PERFORMANCE (GO-AROUND)

ATTITUDE	POWER		PERFORMANCE	
	MAP/Landing Gear	RPM	Airspeed (mph)	Vertical Speed
	FULL POWER GEAR DOWN FULL FLAPS		80	MAXIMUM

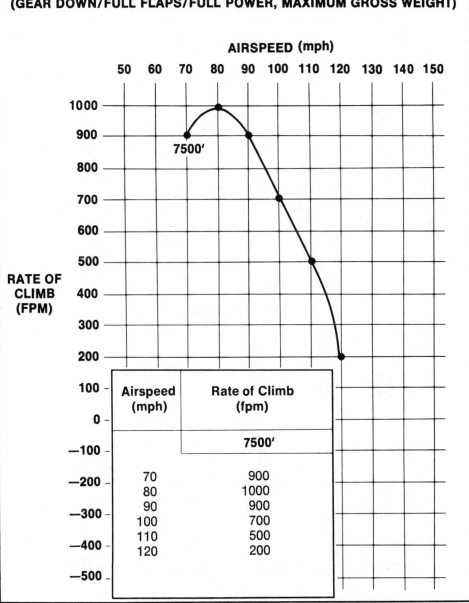

CLIMB PERFORMANCE AT VARIOUS AIRSPEEDS & DENSITY ALTITUDES IN GO-AROUND CONFIGURATION
(GEAR DOWN/FULL FLAPS/FULL POWER, MAXIMUM GROSS WEIGHT)

Airspeed (mph)	Rate of Climb (fpm)
	7500'
70	900
80	1000
90	900
100	700
110	500
120	200

6

MOONEY STATESMAN   180 HP

1

ATTITUDE	POWER		PERFORMANCE	
	MAP/Landing Gear	RPM	Airspeed (knots)	Vertical Speed (fpm)

THE NUMBERS FOR BASIC IFR

ATTITUDE	MAP/Landing Gear	RPM	Airspeed (knots)	Vertical Speed (fpm)
	23″	2400	100	500 ↑
	18″	2400	100	0
	18″ + Gear	2400	100	500 ↓
	23″ + Gear	2400	100	0
	23″	2400	130	0
	18″	2400	130	500 ↓

ATTITUDE	POWER		PERFORMANCE		**2**
	MAP/Landing Gear	RPM	Airspeed (knots)	Vertical Speed (fpm)
THE NUMBERS FOR HIGH-SPEED DESCENT				
	23″	2400	145	500 ↓
	18″	2400	145	1000 ↓

3

LANDING GEAR DRAG AT VARIOUS AIRSPEEDS	
Airspeed (knots)	*Change* in Vertical Speed (fpm)
60	0
70	100
80	200
90	300
100	500
110	700

MOONEY STATESMAN   180 HP

4

MAXIMUM SAFE BANK ANGLE DURING GO-AROUND	
Angle of Bank	**Safety Margin***
0°	20 knots
15°	15 knots
30°	10 knots
45°	0 knots

* This represents the difference between stall speed with gear down/full flaps/full power, and the optimum go-around airspeed of 70 knots.

5

THE NUMBERS FOR MAXIMUM PERFORMANCE (GO-AROUND)				
ATTITUDE	**POWER**		**PERFORMANCE**	
SEA LEVEL	**MAP/Landing Gear**	**RPM**	**Airspeed (knots)**	**Vertical Speed**
	FULL POWER GEAR DOWN FULL FLAPS		70	MAXIMUM
6500'				

CLIMB PERFORMANCE AT VARIOUS AIRSPEEDS & DENSITY ALTITUDES IN GO-AROUND CONFIGURATION
(GEAR DOWN/FULL FLAPS/FULL POWER, MAXIMUM GROSS WEIGHT)

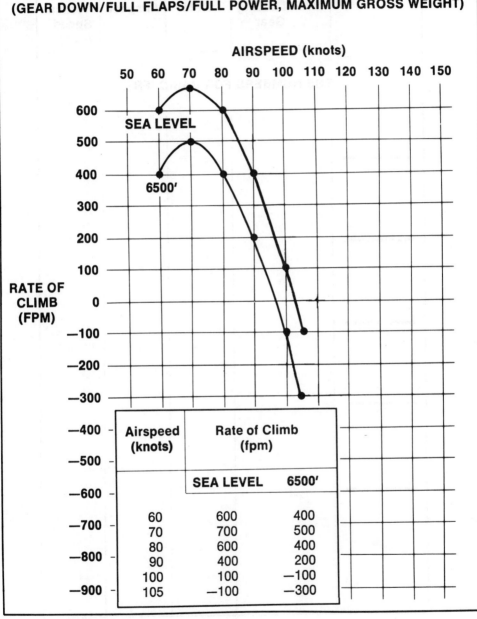

Airspeed (knots)	Rate of Climb (fpm)	
	SEA LEVEL	6500'
60	600	400
70	700	500
80	600	400
90	400	200
100	100	−100
105	−100	−300

1

ATTITUDE	POWER		PERFORMANCE	
	MAP/Landing Gear	RPM	Airspeed	Vertical Speed (fpm)

THE NUMBERS FOR BASIC IFR

ATTITUDE	POWER		PERFORMANCE	
	MAP/Landing Gear	RPM	Airspeed	Vertical Speed (fpm)
THE NUMBERS FOR HIGH-SPEED DESCENT				
———				
———				

2

LANDING GEAR DRAG AT VARIOUS AIRSPEEDS	
Airspeed	*Change* in Vertical Speed (fpm)

3

213

4

MAXIMUM SAFE BANK ANGLE DURING GO-AROUND	
Angle of Bank	**Safety Margin***

* This represents the difference between stall speed with gear down/full flaps/full power, and the optimum go-around airspeed of   mph/knots.

5

THE NUMBERS FOR MAXIMUM PERFORMANCE (GO-AROUND)				
ATTITUDE	**POWER**		**PERFORMANCE**	
	MAP/Landing Gear	**RPM**	**Airspeed**	**Vertical Speed**
_____	**FULL POWER GEAR DOWN FULL FLAPS**			**MAXIMUM**

CLIMB PERFORMANCE AT VARIOUS AIRSPEEDS & DENSITY ALTITUDES IN GO-AROUND CONFIGURATION
(GEAR DOWN/FULL FLAPS/FULL POWER, MAXIMUM GROSS WEIGHT)

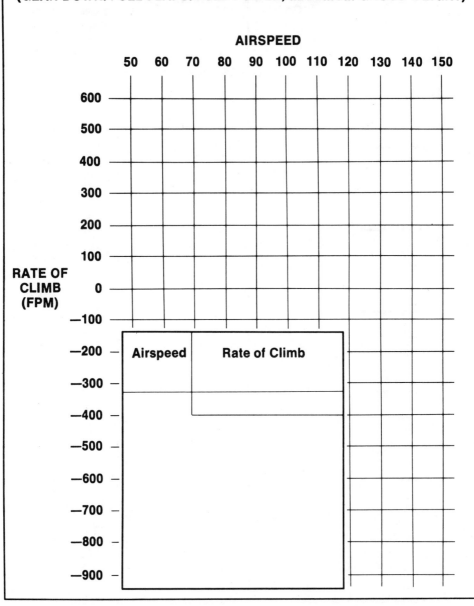

1

ATTITUDE	POWER		PERFORMANCE	
	MAP/Landing Gear	RPM	Airspeed	Vertical Speed (fpm)

THE NUMBERS FOR BASIC IFR

ATTITUDE	POWER		PERFORMANCE	
	MAP/Landing Gear	RPM	Airspeed	Vertical Speed (fpm)
THE NUMBERS FOR HIGH-SPEED DESCENT				
———				
———				

2

3

LANDING GEAR DRAG AT VARIOUS AIRSPEEDS	
Airspeed	*Change* in Vertical Speed (fpm)

217

4

MAXIMUM SAFE BANK ANGLE DURING GO-AROUND	
Angle of Bank	**Safety Margin***

* This represents the difference between stall speed with gear down/full flaps/full power, and the optimum go-around airspeed of   mph/knots.

5

THE NUMBERS FOR MAXIMUM PERFORMANCE (GO-AROUND)				
ATTITUDE	**POWER**		**PERFORMANCE**	
	MAP/Landing Gear	**RPM**	**Airspeed**	**Vertical Speed**
——	FULL POWER GEAR DOWN FULL FLAPS			MAXIMUM

CLIMB PERFORMANCE AT VARIOUS AIRSPEEDS & DENSITY ALTITUDES IN GO-AROUND CONFIGURATION
(GEAR DOWN/FULL FLAPS/FULL POWER, MAXIMUM GROSS WEIGHT)

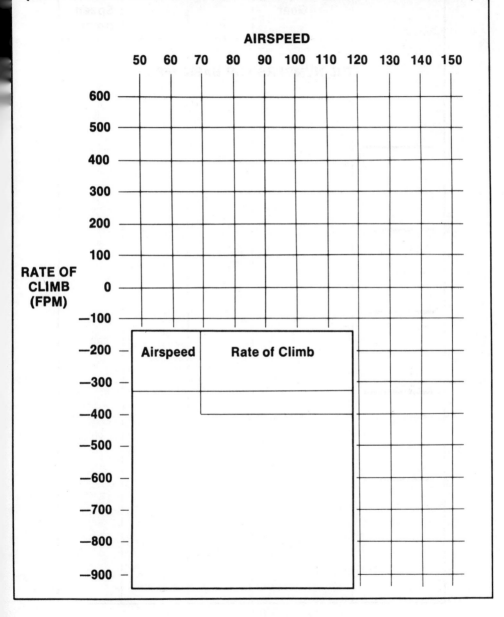

1

ATTITUDE	POWER		PERFORMANCE	
	MAP/Landing Gear	RPM	Airspeed	Vertical Speed (fpm)

THE NUMBERS FOR BASIC IFR

ATTITUDE	POWER		PERFORMANCE	
	MAP/Landing Gear	RPM	Airspeed	Vertical Speed (fpm)
THE NUMBERS FOR HIGH-SPEED DESCENT				
—				
—				

2

LANDING GEAR DRAG AT VARIOUS AIRSPEEDS	
Airspeed	*Change* in Vertical Speed (fpm)

3

221

4

MAXIMUM SAFE BANK ANGLE DURING GO-AROUND	
Angle of Bank	**Safety Margin***

* This represents the difference between stall speed with gear down/full flaps/full power, and the optimum go-around airspeed of   mph/knots.

5

THE NUMBERS FOR MAXIMUM PERFORMANCE (GO-AROUND)				
ATTITUDE	**POWER**		**PERFORMANCE**	
	MAP/Landing Gear	**RPM**	**Airspeed**	**Vertical Speed**
——	FULL POWER GEAR DOWN FULL FLAPS			MAXIMUM

CLIMB PERFORMANCE AT VARIOUS AIRSPEEDS & DENSITY ALTITUDES IN GO-AROUND CONFIGURATION
(GEAR DOWN/FULL FLAPS/FULL POWER, MAXIMUM GROSS WEIGHT)

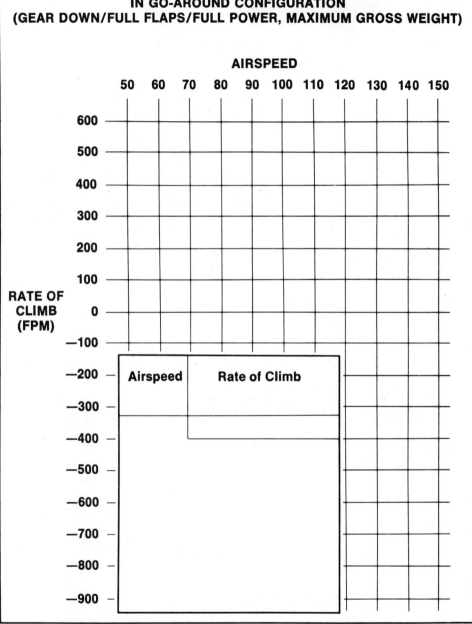

INDEX